Homesteading Adventures
A Guide for Doers and Dreamers

Also by Sue Robishaw

The Last Lamp, a novel
Carlos's ma's Friends, a novella
Frost Dancing, Tips from a Northern Gardener
Rosita and Sian Search for a Great Work of Art

Homesteading Adventures
A Guide for Doers and Dreamers

by

Sue Robishaw

Illustrations
by
Steve Schmeck

Foreword by Jd Belanger

ManyTracks

Library of Congress Catalog Card Number 97-72380

ISBN 0-9652036-1-1

10 9 8 7 6 5 4 3 2   03 02 01 00 99

Printed in the U.S.A. on 100% Recycled Acid-free Paper

Second Printing

Robishaw, Sue (Susan J.)   1952 -
Homesteading Adventures, A Guide for Doers and Dreamers

Bibliography
Includes index.
1. Country Life - Handbooks, manuals, etc.
2. Organic gardening
3. Earth sheltered houses
4. Solar energy

S521 R   630.1   97-72380

Portions of this book have appeared previously in various
forms in the pages of *Countryside* magazine.

Dedication

This book is dedicated to the many homesteaders, past, present, and future, who have found joy in doing, satisfaction in sharing, fun in learning, and life full of living.

Contents

Foreword

by Jd Belanger
editor and publisher of *Countryside* magazine

Most of the few people who know a little about modern homesteading have several preconceived notions about the lifestyle. They assume that homesteaders are people who have left good jobs in the city to pursue simple subsistence living by carving a Swiss Family Robinson sort of existence out of the wilderness.

In the majority of cases, those assumptions are way off the mark. Most modern homesteaders have almost boringly "normal" homes, jobs and lives.

On the other hand, with Steve Schmeck and Sue Robishaw, those assumptions come awfully close to the truth.

Perhaps this is one reason so many readers have expressed unabashed envy of this couple after reading Sue's articles in *Countryside* magazine. It's probably why some have likened them to Scott and Helen Nearing, those icons of modern homesteading. And it's certainly one of several reasons why Sue Robishaw's writing is so enthralling.

Her narratives have the same quality that makes us enjoy a good novel or movie. We come away with a fresh perspective, a renewed and revitalized spirit, and a feeling that for a brief time we have been in another place and time where everything fits together much more neatly than in our own cluttered, uneventful, and perhaps meaningless lives. We might not want to imitate these people exactly, but we can't help but admire them!

In spite of this "storybook" quality, Sue's writing is not only entertaining, but informative. She not only describes their homesteading projects, but details how anyone can accomplish them.

In reality, no two homesteads, or homesteaders, are the same. One of the most fascinating aspects of the lifestyle is its infinity of options and variations. Contrary to what many people think, homesteading isn't defined by certain activities, nor does it follow any specific rules. (Steve and Sue exemplify this, too).

Because of this variety, there are not books on "how to homestead". Books can only show how certain *individuals* homestead, or explain how they accomplish selected undertakings.

However, there is a more elusive, and perhaps a more important and universal dimension to homesteading that is often difficult to describe or convey to others. That is the *way of thinking* that motivates and sustains homesteaders. Some say it defines homesteading itself. Sue expresses this wonderfully. However, if you think homesteading is simply *doing* things you might miss this aspect, so be sure to watch for it and notice how it melds into the actions.

Actually, that mindset might be what this whole book is about. Yes, the book describes one couple's homesteading adventures, and yes, it explains how to tackle many homestead enterprises. But most of all it demonstrates the mode of thinking that makes it all possible as well as *necessary* for living the good life.

You don't have to live in the woods to acquire and benefit from that. If JJ and CindyLou can do it, so can you.

Preface

I discovered JJ and CindyLou one autumn afternoon meandering through my mind. At the start they were two rather nondescript creatures with whom I began having regular rambling many-tracked conversations. They helped me pick and thresh the dry bean pods with much oohing and aahhing over the varied colors while wondering why I spent so much time keeping all the varieties separate, and wouldn't it be easier to just throw them all in together, it's getting late after all you know, and when are we going to make the bean soup? They supervised the pulling of weeds while speculating on what would happen to the garden if I didn't, and why didn't I just get that big old red tiller going and dig it all in, and wouldn't it have been easier if I had put more mulch on in the first place?

They popped up amongst the apple peels and cores of applesauce making wondering if I really needed to peel the apples, and why didn't I dry the crop instead of canning it, but would it taste as good and would it take less time and what kind of apples were those anyway? And did I really mean to drop that core into the cooking kettle instead of the compost bucket?

Over time JJ and CindyLou came into focus as the neophyte homesteaders you will meet through these pages. Fictional neighbors and friends, they have come through my life bringing fun, amazement, organization and chaos. By sharing our own homesteading experiences with them they've shared with me a new outlook and a different perspective. Though certainly of a fictional nature and not based on any real person or persons they have a tendency to reflect back at oneself ones own self. The good, the bad and the ugly. But more important, the funny.

Even if you can't imagine getting excited about digging around in the wet cold mud for potatoes maybe you can come along anyway and laugh with us. Share in our adventures as we find treasures in the mundane and high scholarship in the everyday. Maybe catch a bit of what we've learned, as we catch a bit of the homesteading experience through the eyes and perceptions of the ever lounging simple speaking long lanked JJ and his enthusiastic song filled petite partner CindyLou.

We've had fun with it all. And I hope you will too.

Acknowledgments

A number of people have helped to make this book happen, including the many people with whom we have had discussions over the years. Too many to name, and many whose names I don't know, they have helped inspire me to write and share, whether by their questions or their answers. It is a give and take for which we are grateful.

A question often asked of us is "How do your families feel about your lifestyle?" I am thankful to be able to say that they have from the beginning been encouraging and helpful, sometimes with physical presence, often with words, usually with understanding. That psychological boost has been, and is, appreciated.

In the early years we looked heavily to other homesteaders, experienced or not, for practical help and inspiration, mainly through books and magazines. We are thankful for all of that help, and thank the many people who lived, wrote about, and shared their experiences.

There is one who stands out for having a particular influence, and he is *Countryside* editor and publisher Jd Belanger. This book probably would not have materialized were it not for his support and encouragement through many years.

Many people have freely shared their knowledge and experiences with us over the years, often becoming friends, and much of that learning is reflected in our homestead, and in this book. Two of those, Larisa Walk and Bob Dahse, allowed me to use their words, and experiences, in the chapter on solar ovens, and I thank them for that.

Most of the photographs were taken by Steve or I. The following pictures were taken by photographer Dan White of Manistique, Michigan: Steve at the lathe, the inside shot of the greenhouse, and the early spring photograph of the house. Several of these pictures appeared originally in *Home Power* magazine June/July 1993. Steve did all of the illustrations and graphic work in the book.

Though I am the main speaker through the following pages the life behind this book is definitely a joint project between Steve and I. We are two separate individuals but our lives are strongly intertwined, and mine would not be what it is today were it not for him.

Introduction

Steve and I started our homesteading adventures some twenty years ago, even before our move north to the backwoods of Michigan's Upper Peninsula. The satisfactions and challenges have followed us unabated since. We find ourselves living both in and out of the mainstream and it is a good way to live.

A homesteading lifestyle is as varied and individualistic as the people who embrace it, and we are no different. I offer our experiences not as rules or dogma to live by but as a picture from which to borrow pieces, to add to a corner of your own creation, to select as a focus for inspiration and ideas.

In the following pages I have shared practical details and wider overviews, actual experiences and philosophical thoughts; often through the humorous conversations with and between the fictional JJ and CindyLou. Though it would be the ideal it is, unfortunately, not practical or possible to talk and visit with everyone personally. But it is my hope that through this book we can continue the tradition that is so much a part of the homesteading world; the tradition of passing along experiences, skills, and knowledge gleaned from our years of living the 'simple' life.

The information presented here, and prices where mentioned, are current to when the book was written. But we are participating in an ever changing world. While many of the ideas and suggestions are timeless, others change with the passing moments and years. Consider this a beginning, not a final statement, and get more information when necessary, or needed. There is much source information in the bibliography.

Since I am the author of the following story it may appear at times as if I am homesteading by myself. But that is not the case. Steve and I are partners in this adventure. Though there are projects and work we do by ourselves, generally we share the load and the fun, both in work and in play.

one

Homestead Dreams

~ ~ ~

"JJ, I have an important announcement to make. Are you ready?"

"How can I be ready when I don't know what you're going to say, CindyLou? I'm not a mind reader you know. And if I was it would be pretty darn uncomfortable around here. It's bad enough . . ."

"This is IMPORTANT, JJ! This is going to cause a great affect upon our lives."

"Well, I suppose I ought to know what it is then. Speech away, CindyLou."

"We are going to adventure out into the wilderness, establish a homestead for ourselves and live self-sufficiently!"

"Now CindyLou, first of all you can't do that anymore. There's no more wilderness and they don't allow you to homestead nowadays, and nobody can be sufficient unto themselves in this day and age anyway. Don't you read the newspapers? Besides, you don't have money to travel to any wilderness, assuming there was any, which there isn't, and if there was it'd be off limits to you, and probably something wrong and dangerous in there anyway, and . . ."

"This is MY dream, JJ, and I will make it whatever I want. And it will do what I want it to do."

"OK, OK, unruffle. You've been talking to Sue again haven't you?"

"Look at this magazine that came in Sue and Steve's mail. It is about people who are doing the very thing I am talking about. They are all having their own wilderness dreams adventures right now, all over the country. All over the world! So there is no reason why you and I can not do it too. Here, read it. But do not take too long. I want to get started today."

"Do Steve and Sue know you been swiping their mail, CindyLou?"

"I did not 'swipe' it. I simply picked up their mail for them. You are the one who is forever harping at me to expand my mind by reading."

"OK, OK. So what do you have there? Whoa, did you take a look at that critter on the cover? Those sharp pointed weapons on his head look mighty dangerous to me. Think of the holes that guy could make in your mohair sweater, not to mention your body, and you thought that mouse was a problem. And it's not like we need more problems. Why . . ."

"Read the words, JJ. And those are not weapons, those are the cattle's horns. Have you never seen a cattle before? This is an important crossroads in our lives, JJ, be serious! We do not have to have a . . . whatever kind of cattle that is, to be a homesteader. At least I do not think we do. I do rather hope we do not. He does look big. And he must shed something terrible."

"Well, I myself can't quite see cozying up to a critter than could make pancakes out of parts of you with one sidestep. And when did you ever see a cattle anyway? Besides, I'm getting to feeling pretty comfortable here and I'm not of a mind to go gallivanting around looking for a place for you to have your adventure."

"We, JJ, we. And we do not have to go anywhere. This is the New Homesteading. It is a lot like the homesteading in the past but now we are thinking wider and higher. We will homestead where we are, or where we can. Right here. This is our new homestead. This is our wilderness. This is where our dream begins. Behold!"

"I behold a farmed out old field with a fallen down old shed splatted down in the middle of it. And if you behold that shed too close you're going to be looking like a pincushion full of nails. And that's not a pleasant prospect. A rusty old nail could cause all sorts of . . ."

"Look at those trees over there, JJ. They may not be big yet but they will grow tall and strong given time and care, which we will provide. You have to look wider, JJ. That old shed is a gold mine of valuable and usable lumber."

"And mostly kindling. Not to mention work, and time and sweat and mosquito bites and black fly chomps and . . ."

"So? What have you better to do, JJ?"

"Me? Now see here, CindyLou, don't be dragging me into all this. Besides, what's the use? You can't live like that for long and why do all that work just for a lark? There's not none of us getting any younger you know."

"Age is a state of mind, JJ. You can be as fit as you allow yourself. And it is past time that we both allowed ourselves some good physical exercise. And this is how we will get it. Look at Sue and Steve. They live 'that way' right now and they are very happy at it."

"Sure, but they're different."

"Well, so are we. Do you know of anyone who is *not* different?"

"There are lots of folks who don't want to be different, CindyLou."

"Well, I do. And I want to find out more about this living with the Earth."

"Can't argue with that I guess, CindyLou. Though I was thinking that we're living a mite closer to the Earth right now than I'd planned. And if you get any closer to the Earth, CindyLou, I won't be able to see you. Sorry, that just slipped out. Unruffle, unruffle, I apologize. So, where do you plan on starting? No matter what, you can't just go off and build yourself a wilderness dreams adventurous homestead life, or whatever it was you said you were doing, without doing some good research first. You have to know where to start. And I can't see starting anything in this old tent except claustrophobia. Now if you hadn't had that little accident with your Mom's reeeeallly nice and reeeeallly comfortable and reeeeallly spacious RV . . ."

"We, JJ, we. We are going to adventure into a bigger and better life. And you do not need to bring up the RV. And I do not want to spend great big chunks of time 'researching'. If I let you get going in the library we will *never* have a homestead. We are going to do it right now."

"Now, CindyLou, you know you have to think first. Dreams are OK and all but you know what happens when you just take off and do something without thinking or researching some first. Remember the hair color of the week scheme . . ."

"There is no reason to bring up events that are past, JJ. I will consider *some* research first, as long as it does not take too long. But where to start I admit I do not know."

"Well, it seems to me we might as well go talk with Steve and Sue. As you said, they've been doing this crazy thing for quite a time now. Then again, I'm thinking they might be a bad influence on you. They might be happy and all today, but how about tomorrow? Of course, who can even imagine what tomorrow will bring, if there is even such a thing as tomorrow. Did you hear on the news last night . . ."

"A place to live. That is where we will start. We will build us a home, JJ. It does not matter if tomorrow decides to come or not. We will still be, and I wish us to be on our homestead. So that is settled. Come along, JJ. *Ohhhh, we'll find us a homestead, we'll laugh in the winnnnndd, find treasures in sunbeams, and life will begiiiiiiiiiinnn.*"

"For heaven's sake, CindyLou, volume down, volume down. You're waking the worms!"

"The worms are already awake, JJ. In fact, they were singing harmony. You need to listen more, JJ. If you are not going to join in the singing, then you should join in the listening. 'See you later, future JJ and CindyLou Homestead. We will be back to meet all of you future neighbors soon.' We do have to consider the others living here, too, JJ."

"Speaking of neighbors, isn't this spot here sort of real close to Steve and Sue's homestead? I thought they said they were happy not having neighbors real close? Have you told them you're thinking of moving in here permanently? "

"Well, we will have to tell them of course. But we can do that later. We will slip it in at just the right moment."

"Seems like we'd be better neighbors if we didn't try to slip anything anywhere at anytime, CindyLou. That's what's so wrong with the world today. Too many folks slipping things in when they think no one is looking. Only there's always *somebody* looking, and then things get worse, and . . "

"Then we will choose another spot. Now let us go, JJ. Arise your body to a gentle jog anyway. We are going to have to be in very good shape for our homestead adventure you know. And we have a very lot to do."

"I don't think I like the sounds of that, CindyLou, this is supposed to be a *vacation*. Oh OK, you jog on ahead, I'll be along. My teeth can't handle the rattle involved in jogging. Neither can my toes. Toes don't last forever you know. And healthy teeth are important for any number of functions, not the least of which . . "

~ ~ ~ *two* ~ ~ ~

Beginnings and Bugs

~ ~ ~

"Hello, Sue, are you busy?"

"Hi there, CindyLou. I'm not so busy I can't take a break to talk with you. Isn't it a great Spring day? Just listen to all those birds. So, what's up, you look like you're excited about something."

"I am, I am. JJ and I are about to embark on an adventure. We are going to create for ourselves an earth friendly homestead. We will take ideas from the past and the present, the future and our dreams. Combine them all to make our own life right now! Or as soon as JJ gets here anyway."

"Mmm, sounds vaguely familiar. That's quite an undertaking, CindyLou. Hi there, JJ! That's OK, sit down here and catch your breath. Could I interest either of you in a drink of water?"

"That would be very nice. JJ and I are getting in shape for our new adventure. Drinking good, fresh water is an important part of that program."

"Can't a body just have some because they're thirsty? This jogging takes a lot out of a person you know, and I haven't a whole lot to spare."

"It is good for you, JJ. By the way, your rhubarb is not by any chance up yet is it, Sue?"

"CindyLou! That's begging!"

"I am not! I am just . . ."

"It's OK, JJ. The rhubarb and I would be happy to share with you and CindyLou. I wouldn't mind having a bite myself. Let's sit down under this apple tree and you can tell me all about your new adventure.

"You two both have enough shade? You might want to grab some of that fresher hay there to sit on CindyLou, we wouldn't want to dull that, um, nice bright pink of your slacks with streaks of rotting hay. These early sunny days sure feel good but they are warm."

"Yeah, and it's nice not to have to bring along the bow and arrows to fight the mosquitoes and black flies. I heard they're coming though, real soon. They're just waiting for us to get nice and warmed up first. Easier to bite into then, spreading disease, destruction and mayhem, not to mention . . ."

"JJ, they can not be that bad. I do not believe that they can be. Are they Sue? Besides, I have not seen one yet."

"Well, they *can* be pretty irritating, some years are worse than others. But you'll get used to them, CindyLou. You pretty much have to if you're thinking of staying in this area. You learn to live with them. And if we didn't have mosquitoes and black flies we wouldn't have swallows and bluebirds. It's worth it. So, where are you on your homestead plans?"

"Nowhere. We haven't even done any research yet,

and I could tell you stories about what happens when CindyLou goes off on a project without thinking it through first!"

"But I *am* researching, JJ, we are here are we not? We are rather at the beginning of our homestead adventure. Would you tell us how you and Steve started? Then we will know what to do next. I want to be settled into my new homestead before it gets too hot this summer."

"Well, CindyLou, it might take you a bit longer than that, depending on what you have in mind. And it won't be the same for you two as it was for us, of course. But I'll be happy to share our experiences. Maybe it will give you some ideas. Have another rhubarb and stretch out.

This is an ever changing story. The world is changing, the Earth is changing, and we are too. Our homesteading lifestyle reflects all of that. But this is how it all started for us . . .

Steve and I were seated on his mother's couch talking. I think it was afternoon. It was 1975 or so.

'I plan to move to the Upper Peninsula,' Steve told me, sort of out of the blue.

'Sounds good to me,' I replied. I hadn't thought about it before, but why not?"

"You mean that's it? You just agreed? You just decided to change your life and take off without thinking about it or doing any research or . . ."

"Not exactly JJ. I'm coming to that. But at that time it was one of those ideas that simply felt very comfortable. Almost as if I already knew it but just hadn't brought it into my mind yet. I don't recall thinking about it that much. Once the idea was there it seemed a given. If it makes you feel any better, just because I hadn't thought of it before doesn't mean Steve hadn't. It had been on his mind in some form or another most of his life. Besides, adventures have to start somewhere."

"That is what I have been trying to tell you, JJ. We

cannot wait for you to research this all thoroughly to your satisfaction before we continue with our plans. We have to go *do* it."

"Wait just a minute, CindyLou, that isn't exactly what I'm saying here. That was just the very beginning. How about if I continue. There is a bit more to it than that."

"All right."

"Fine by me."

"Our homesteading adventure was born and it grew. We subscribed to *Organic Gardening and Farming.* Friends introduced us to John Shuttleworth's *Mother Earth News.* We bought homesteading books, how-to books. We made lists of what we wanted to take with us. We became avid auction goers. We went, and bid and collected our 'homestead'. Our city home became merely a storage shed.

Somewhere in there we headed north to look for our future homestead. We assumed it would be on a lake or river, or at least have a creek running through. We didn't waste time mourning that ideal when confronted with the price of such land. We visited real estate offices, walked across 40's and 80's and 160's. Found vultures and Lady Slippers and interesting looking small insects that we picked off ourselves by the dozens after one excursion. Later we learned their name. And understood belatedly why a real estate fellow was so aggressive in smashing the one Steve casually picked off his shirt, "squashed" between his fingers and put in the man's ashtray. At the time we wondered about the man's aggressiveness towards such a small insect. Without knowing it, we had just met the later to be common acquaintance, the wood tick.

Within two weeks we found our dream. Eighty acres, seasonally accessible, cut over woods and farmed out fields, no buildings, no near neighbors. Such potential.

Our 5 year plan was to work in the city, pay off the

land contract, work on the homestead on weekends (it was 300 miles away) and move when done. We bought fruit trees, were introduced to *Countryside/Small Stock Journal* by the nursery owners, tasted goats milk for the first time thanks to the same folks. A home site was picked out, then another. Fence posts and fencing went up, fruit trees in, plans made. It was going to be quite a homestead. While still in the city we learned to bake our own bread and grind our own flour. We created plans and drew ideas. We talked and thought and read and had a very good time.

Two years later we couldn't stand it. We quit good paying jobs, left the city and, with the patient help of Steve's sister and brother-in-law, moved to the Upper Peninsula, in May. We all found out about black flies. And U.P. mosquitoes. We discovered headnets. And learned not to invite friends up in the spring if you want them to ever return.

Our new home was a 6 ft x 8 ft camper cap installed on a plywood base. Real cozy. It was great. It was paid for. We soon found a restaurant in the nearby town (15

miles away) that would let us sit in a corner booth talking and planning and drawing on the place mats for hours on rainy days. I learned to drink coffee.

We planned that first summer to get the windmill installed (build the tower first), work up a garden, get a good start on building our house, build a shed, get two years worth of firewood in, clear some land, work on the road and the driveway. We'd done our homework. We knew what you had to do to have a real homestead. Well, maybe it would take us two years to get it all done, including the barn and pasture renovation. And should we have a cow or goats? Chickens of course. Sheep? Pigs?

By fall we had the windmill up, tower and all, over the newly drilled well. Beautiful structure, beautiful sound, beautiful water. And that in spite of Steve having to deal with a wrenched back in the middle of it all. We had the land excavated for the house. We dug footings and had them poured. Muscles appeared that we hadn't even guessed were possible. We put in a garden. The grasshoppers loved it. We were brown from hard work and sun. It was quite an accomplishment. We know that now. Back then we just thought of all we *hadn't* done. Our list was as long as anyone's.

We had enthusiasm. We had no money.

Back to the city. We lasted two months. We decided we'd rather be poor on our homestead than making money in the city. Our ignorance and desire sustained us. We made the decision Thanksgiving weekend. We were going to borrow $1500 for wood to build a cabin. The night before we left, Steve's uncle brought over a check for $1500, Steve's share of his grandmother's estate which he had not expected. We accepted this encouragement and moved back north.

Between Thanksgiving and Christmas we built the cabin (temporary of course, we would for *sure* get our house built next summer). We had a small enameled garbage burner in our small plywood and camper cap home. It was over 100 in the cab-over bed area, below freezing on the floor. We slept in four hour shifts since someone had to keep feeding the burner semi-hourly. It was cold. It snowed. We loved it. We learned to snowshoe. And to cut firewood on snowshoes. And to turn around and backup on snowshoes without falling down into the deep cold white. And we learned to get up when we did, even while laughing like fools. We learned how to work and live in the cold and snow.

We moved into the cabin before Christmas. Most of the walls were only foam board as yet, the one window polyethylene, the door a blanket. And there was frost on the nailheads on the subfloor. But it had a new steel plate stove that would not only blast you out of the cabin with the heat but would keep a fire all night. Heaven! Some joys are hard to explain. This was one such.

We went out only once or twice a month that winter. It snowed almost daily. Great drifts. We snowshoed everywhere, sometimes even to the outhouse. We worked on our cabin. Our land had once a good number of elm trees in its forests. They had all died before we got there. We were sorry for that, but grateful for the dried-on-the-stump firewood. It provided us with much needed heat and cooking fuel that winter (and for almost 15 more winters). We cut wood every day or two, whenever the weather permitted, hauling it on inadequate, but available, sleds. A neighbor often called to let us know if a large storm was expected and we would haul in extra wood.

I learned about boots that were too tight, and toes that never seemed to completely thaw. Steve learned to keep his snowshoes out of the way of the chainsaw. Wire wrapped around a nicked tip made a "temporary" patch that is still there eighteen years later. In the future we would learn all too well the much joked of, but all too certain phenomena of "temporary" homestead repairs and projects.

We reveled in the beautiful wood around us. A wild

black cherry tree that had been cut down to make way for our cabin provided us not only with firewood, but carving wood as well. Steve taught me the art of sculpting wood. We made hand carved wooden spoons. We started to recover from the years of city life.

Time went by. We haven't had a winter like that one since. But we did get our house built, eventually, moving in some six years later.

Over the years, to our occasional amazement, some of the things on the old list did get done, more later than sooner. Many more things were crossed off over time (thank goodness). And many things done are being redone now. We are still doing but aren't making too many lists. And if something on the list doesn't get done this year, it will next, or maybe in five years. Or maybe it just wasn't that important. And that's OK. A lesson long in coming. We caused ourselves a lot of grief in years past by our artificial *have to's*, but we've had a lot of fun too.

We play sometimes with the idea "if we were to start over what would we do" and yes, there is a lot we'd do differently. But we were where we were then, and you can only start from where you are. I can't say I'm sorry for anything we did. We learned. I can't believe it was so hard to learn to slow down, to learn to live and laugh and play. Or rather I can't believe I made it so hard. Make it so hard yet sometimes.

Our philosophy is gentler now than it was then and closer to the Earth, so our ideas and thoughts now reflect that. We no longer think so much in terms of goals to achieve as we strive to just enjoy doing whatever it is we are doing.

JJ, CindyLou, you two awake there?"

"Of course I am awake, and thank you for the story. But I did not have in mind that I would have to move so far away in order to begin our adventure."

"You don't have to move, CindyLou. You can change your life right where you are. For us, moving north was a part of it. It was in the days of the 'back to the land' movement, and many folks were moving, trying to find a better place to live, mostly moving out of the cities and suburbs. But it soon became obvious to us that our life on the homestead wasn't any kind of movement, it was just a way of life. It doesn't matter where you are. What matters is what you do with what you have."

"That would make a good song, *It doesn't matter where you are, it's what you do with what you havvvvvve . . .*"

"You don't *have* to sing so loud, CindyLou. You're going to wake up every munching cutworm in the area! And there won't be anything left in the garden when they're done and we'll all starve."

"It does not work that way, JJ, and you know it. It is a very good song and if one feels like singing they should."

"Well, we're supposed to be here to get answers so you can make your homestead, not singing at the world. Besides, the world's so full of problems I don't know what you find to sing about. Though I do think lying here in the sun contemplating life is a very good solution. And it's not as if there is any rush. The world is going to fall apart whether we do anything or not. Yep, I think we ought to just lie around like this and think about it. Quietly. Sue's not going to invite us back if you warble at her every time we come."

"I enjoy CindyLou's singing, JJ. And sometimes it's the best solution for a problem! Now, I don't pretend to have all the answers, or many even, to the problems and griefs of today's world. And our way of life isn't an answer either. But the world *is* changing, just as the weather patterns and storms are, and I think things are speeding up. There doesn't seem to be as much time to wishy-wash around, think things over, procrastinate, daydream your intentions away. I think we all have to begin living right now the way we dream of living someday, when we have more time, or more money, or more experience, or feel more like it. It's important to plan and think and research your project. But it is also important to get going on it. You don't have to do it all at once you know."

"I suppose. But you have to start somewhere and some somewheres are better than others. CindyLou wants to grab a hammer and start throwing up a house. But if we don't learn some first it won't last through the first snow. Besides, we've never build anything bigger than a bird feeder, which may not last through the next rain. I tell you, CindyLou, you just can't do it. It doesn't matter what you've read. You just can't go building your own . . . Now, come on back here . . . CindyLou! . . . It's too nice a day to be hurrying about anything let along *building* something."

"I don't think that was quite the right approach, JJ, telling CindyLou she couldn't do it. Besides, I think you two can. We did. And we hadn't much more experience than you two when we started."

"I don't know, maybe. I kind of wish you hadn't shown CindyLou how she could saw a board off and hammer a nail into it. And even if we do decide to build something we have to have some plans first."

"Maybe if I tell you about building our cabin, and the house, that will give you some idea of where you're headed. And we have a lot of construction and building books in our library that you can borrow. Dreams, plans, and implementation are all important. But it works better if you get them in the right order. By the way, where are you planning to build?"

"Well, umm, I guess CindyLou has some ideas on that, I'm not too particular myself. But, well . . . You know, having neighbors near by could be real handy. Say you're making a cherry pie and you run out of sugar. There you are, miles from a store, all those cherries just waiting to go into the pie, and you with no sugar. What would you do? Now if you had . . ."

"I'd put the cherries into a bowl, pour maple syrup

over them, and eat them. Or, more likely, they'd stay in their bowl on the counter, we'd grab a handful now and then as we go by, and before long they'd be gone. I'm not much for making pies."

"What? Oh, I see, well . . . But now if you had neighbors near by . . . What's that?"

"Sounds like sawing and hammering. Come on, grab a rhubarb for the road and we'll go over and talk with CindyLou. You know, JJ, we don't mind if you and CindyLou decide to build on that land next to ours. It's a nice spot. That shed there behind where you put your tent has some real possibilities. It's nice to start with something, even if it is a bit dilapidated."

"Yeah, well, I suppose so. I mean, I'm glad. Not that we're really going to, you know, it's just an idea of CindyLou's. A person can't just up and do that kind of thing nowadays. Not that you and Steve aren't people or anything. But I just read the other day that . . ."

~ ~ ~ *three* ~ ~ ~

Walls, Roof, and Windows
The Cabin is Born

~ ~ ~

"That's a fine, um, bench? you have going there CindyLou."

"It is not a bench, Sue. It is the beginning of our home. It is a door step."

"Now what are you going to do with a door step, CindyLou? We haven't even got a place planned yet. And you sure don't need a step to get into the tent. If you put that in the doorway I'm sure to break a leg or something trying to get out. And then where will we be? Why just trying to . . ."

"Actually, JJ, that is a pretty nice step CindyLou has built here. You could set it beside your tent and use it to sit on for now. See. Oops. Well, a few shims will take care of that wobble. Ouch! Ah, CindyLou, you might want to move this nail here so it heads more into the wood. But I think it's a very nice start."

"Thank you, Sue. I am sorry about that nail. I just wanted to get started on our homestead adventure right now. Although I do admit I am not sure where to go from here."

"How about taking a break and I'll tell you how our cabin came to be built. Maybe you'll get some ideas for what you would like to have connected to your doorstep. And how to do it."

"Yeah, it's too warm to be working so hard anyway, CindyLou. Let's go sit in the shade and listen about their place. Then we can talk about what you want yours to be like."

"*Ours*, JJ, ours. This is a joint homestead adventure remember. But if we do not work on the nice days we will not have a roof over our heads by the time those horrendous snows I have heard about arrive here."

"I don't think you have to worry about the snow yet, CindyLou. And the winters here really aren't that bad. But let's take one thing at a time. Let's walk back to our place and take a look at our cabin, which is now our workshop, and what started out as our home. I think you'll be happier in the long run if you take time to do a bit of thinking before you start right in building. If you and JJ come up with a plan you'll be able to get a lot more done during these nice days. And enjoy it more too."

"You are probably right. It might be helpful to have more details about what you have built. And I admit I do not know what it is I want to build yet. Sally forth!"

"Hey, watch that saw, CindyLou. It's not like I have extra hair for you to be chopping off. Hair doesn't grow on trees you know!"

"I did not come that close to your hair, JJ. Come on, I want to look at their cabin right away. Maybe that is what we will build. Did you like living in a cabin?"

"Yes I did, I still miss it in a way. Don't get me wrong, I like the house, it has its advantages, and makes the homestead life easier in many ways. Though I'm coming to understand that for every ease that we concoct for ourselves, we pay for it in some other, maybe subtle, way. But the cabin was a good size for us to build for our first project. The construction was relatively simple, and the eventual layout comfy.

To be honest, the building was a bit small to both live and work in. Bulk food storage necessitated creative efforts. You certainly couldn't get too uptight about what a certain room was *supposed* to be used for. It was for whatever you needed at that moment. When you only have one (then later one and a half) rooms this outlook is a necessity. Here is the layout when we were living there:

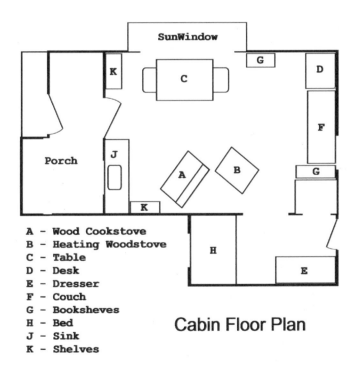

A - Wood Cookstove
B - Heating Woodstove
C - Table
D - Desk
E - Dresser
F - Couch
G - Booksheves
H - Bed
J - Sink
K - Shelves

Cabin Floor Plan

The structure of the cabin, a simple shed roof design, was based in large part on ideas from the book *Low Cost Energy Efficient Shelter*. And on Steve's drawings which he'd been working on since he was eight or nine years old. The main inspiration for those was a small company cabin near AuTrain Falls where his family had spent their Augusts from before Steve was born.

Our cabin was built with green wood, which was what we could afford. And as the wood dried the cracks widened, both in the floor and in the outside siding. We had some pine slabwood we'd gotten from a local mill to cut up for kindling and decided to give those a try for battens on the outside walls. They worked fine and the natural bark edges turned out to be an unexpectedly nice addition to the look of the place. The main room of the structure is 14 ft x 22 ft.

The back room was added on several years after the original structure was built, a birthday present from Steve's Mom. Though only 6 ft x 14 ft it was quite an improvement for living area. We moved our bed from the main room into the addition and added some shelves for needed storage. The area under the bed was our root cellar. Unfortunately, it was also quite shiveringly cold back there in the winter. We discovered the warm wonder of a space blanket placed under the bottom sheet. Ahhh. And I learned to sleep with a wool chuk on my head. Instant understanding about the old fashioned nightcap. Don't laugh, it makes a big difference for comfort when your bedroom is freezing. But, you know, we never slept better.

At the other end of the seasons we faced a different threat to our night's rest. The constantly, and I do mean incessantly constantly, buzzing of the U.P. state bird, the mosquito. It would be fine if they would just bite you then shut up. But I guess that's not their idea of fun. In defense we constructed a lightweight frame around the bed up to the ceiling and attached mosquito netting to it. The side had a mosquito netted door so we could get in and out. One couldn't just jump out of bed without thinking, but it made life, and sleeping, amongst those jungle beasts bearable. Especially the first summers which were particularly prolific with the critters and our cabin particularly accessible to them. Living in the cabin had its unique points. When we had chickens roaming the area they particularly enjoyed the space underneath the building. And would drive us crazy with the incessant pecking at the exposed foam board under the floor. When the cold came we waited eagerly for enough snow to fall so we could shovel it up against the cabin creating a nice cozy bank against the cold drafts. It made for a much warmer floor inside.

Furnishings were at a premium and most things had to express more than one purpose in order to stay around. We were rich with a table and four folding chairs, the bed, and a three legged shaving horse. Then one day early on we were expecting company and realized the seating we could offer was inadequate, unless you counted the floor. So Steve grabbed some leftover floor boards and some scrap two by fours and knocked together a temporary couch which was attached to the west wall. A few quilts thrown over and in less than an hour we had 'regular' furniture. Well, as regular as we get. Later on we added a foam pad under the quilts which made a great improvement for those folks who sat down *expecting* to hit some padding.

[Seven years later we moved into our house with much the same furniture - a table, four folding chairs and an additional two wooden kitchen chairs. We had invited a number of folks over and realized, once again, we were short of seating. We had some large pillows which we spread out on our indoor lumber pile. Instant couch. But we needed more. Steve took hammer and saw to the temporary couch which had been left in the cabin. It was detached from the cabin wall, a few boards added, and soon it sat as a temporary couch in the house living room in time for the visitors.

Eight years passed and the temporary couch again felt the saw and hammer. This time we needed an extra bed. A few more boards, some bolts and nuts and there we were. Instant hide-a-bed. Flipped up it was a couch, set the back down on hinged legs and it was a bed. We were getting so sophisticated it was scary. But I guess it could no longer be called temporary. We spent some time with the file and sandpaper rounding and smoothing, and I oiled it. I think that made it an official piece of furniture and no longer in the temporary class.]

The cabin has a porch which was a fun project over many years. It started as a platform between the cabin and the plywood/camper shelter (which we had first lived in and which was now a storage shed). After a bit the camper (still on it's plywood base) was scooted and shoved and pulled (come-a-longs are great tools) farther away from the cabin. We then added a roof to our porch. That was nice. Then a half wall was built around to keep the rain out. Then glass windows were put in to help keep the cold and snow out in the winter. I guess it would have just continued growing if we hadn't moved our attentions to the house. It was all built with scrap and salvaged materials which is the most satisfying way to build.

The cabin is now a workshop for the woodworking power tools. It is full of wood and sawdust and machines. And usually a number of projects all waiting for attention. The floor plan as a shop has been redrawn many times as our power supplies changed. At one time Steve had a line shaft halfway down the middle of the main room, with tools attached on either side with belts and handmade hardwood pulleys. Some of the tools moved in and out of the line as the needs changed. The power was a Tecumseh 6 hp gasoline engine from our front end tiller. It sat on a stump outside and connected with a pulley to the shaft coming out through the wall to meet it. It worked well, though it was hard to start in the winter.

Then a new tool arrived and things changed, as is a common course of events. We bought a Ryobi 10" planer, and to run it we had to get also a generator. So, a

Kawasaki 2300 watt generator joined the troupe of gasoline hogs. The line shaft was dismantled, the Tecumseh went on to other things (seldom in its life on the tiller it had started on), and the generator ran the power tools. And the washer. And the flour grinder. We traded two spoons for a used vacuum cleaner. Someday we wanted all our tools to run from power produced by the sun and wind. But for the time being, the gasoline generator would do.

Someday the cedar posts on which the cabin was built will probably have to be replaced, not a project we are particularly looking forward to. It would have been nice to have built it originally on stone/cement piers but cedar posts were what we had and could afford at the time. The roll roofing has held up pretty well with an occasional patch job now and then. But shoveling snow off in the winter is hard on it. Ideally in snow country you would build with a steep pitch, or an extra sturdy construction so it can handle the snow load. We didn't understand that when we built. So when the snow gets very deep we shovel the snow from our roofs, including the storage shed and wood shed. It is not a bad job and rather fun to shovel and slide great big slices of snow off to whoosh into the snow below. Then one final not so graceful whoosh "eeiiyyhhya" as the snow pusher can't resist taking that same path off the roof. We are now planning to rebuild all the outbuilding roofs with a steep pitch and metal or cedar shingle roofing.

A more challenging winter problem in the cabin was our water supply. We had buried our main line down to where the house would be and had buried a side line to the cabin, which was fine. But the line from the ground into the cabin was not fine. Though we piled snow and insulation around it as winter progressed it usually froze. And we got to pump our water by hand and carry it the 300 or so feet down to the cabin until spring's warm weather would thaw the water line. We continue to view running water as a luxury which we don't take for granted. We now have a hydrant at the cabin/main water line junction and we could have used one of those up by the cabin (or in it) when living there. When closed the hydrant shuts off the water and drains what is in the hydrant pipe into a dry well below frost line. When you want water you simply open the handle, the drain closes, and the valve opens to let the water up into the pipe again. Nice tool.

We moved into the cabin long before it was finished and while we lived there we never did get much of the interior wood up on the walls and ceiling. We came to like the soft, kraft brown color and texture of the fiberglass insulation on the ceiling. And I found quilts hung on the walls made nice cozy temporary wall coverings. They added a nice bit of color too.

On the ceiling over the bed in the back room I tacked up a patterned sheet. Since there wasn't much room between our bed and the ceiling this made that area a little more comfortable. And when the mice took to running races across the sheet there was some diversion in the game of trying to swat them as they went by. Though I can't say as it ever slowed them down, nor discouraged their visits.

The mice appeared to appreciate the cabin almost as much as we did. And we discovered the "you scratch my back I'll scratch yours", or more precisely, "you scratch my back, a lot, and don't forget the cheeks and chin, and feed me, and I would like more tuna and cheese in the diet please, and I'll catch the mice, when and where I feel like it, when it suits my fancy, and don't be silly of course I eat birds too and it's too bad you want to have a few chipmunks and red squirrels around, they are dinner and fair game, that's part of the deal, do you mind not making so much noise, I'm trying to sleep here" relationship with the cat. One day a scroungy yellow and white tom walked into our lives, whom we felt sorry for so started feeding, though had no intentions of him moving in permanently of course. So much for what we knew. A hefty vet bill later, some food, some loving, and the natty, semisophisticated MisterC took over his part of the arrangement. Though I can't say the mouse traffic over the bed was any less, we were all happier for the relationship which lasted enough years to make him very much a part of our lives.

When he died just a few months before we moved out of the cabin it did not take us long to decide to make a visit to the Humane Society lady in our area. Long haired black with white Ditto and her gray stepbrother, Brandon, joined our homestead to help us move into the house. Unfortunately, Brandon left us to join MisterC within a year. So Ditto, who has as much fur as body, has reigned queen of the castle since. Hunting hard when she does and sleeping and relaxing with equal intensity the rest of the time, with great contentedness.

Until a few years ago, when, to Ditto's utter disgust, a small, extremely lively black and gold and white abandoned kitten found her way here. And Cali moved in to reign terror and laughter upon our home, and devastation upon the small animal population in her territory. With her extra toes, yipping voice, and loving, independent, constantly curious manner she has expanded our world tenfold. And Ditto, though resigned, has not forgiven us.

All of which, whom rather, may have little to do with our cabin adventure. But it would be hard to leave them out of any discussion about the homestead since, as you probably have noticed, they certainly share this place with us. Or we share it with them. I'm never real sure."

~ ~ ~ four ~ ~ ~

Where The Building Begins
Behind And Beyond

~ ~ ~

"Thank you, Sue. Now I know where to begin our homestead adventure. We will start our home this afternoon. That is a very good place to start. Come along, JJ, we have much work to do. Could we by chance borrow another handsaw? I think we will both need to be sawing if we are to finish our home before the mosquitoes and black flies arrive. Which is very important I would say. Do you think we have a few weeks before that happens? I do hope we have that much time. It may take us that long to finish our home. Come on, JJ, let us get going."

"Now just a minute, CindyLou. Aren't you forgetting a few things? Like plans, and materials? Not to mention labor. I especially want to know just who is going to be doing all this labor. And you don't even know exactly where you want to build yet. Let alone what. A cabin that'll be a shop later? Or a shed? Or a sauna? Hey now, that's an idea. I can just feel that hot steam rolling up around these old bones next winter. Ahhhhh. Maybe we should just build a sauna and forget the house. The tent's not all that bad you know."

"The tent is fine, JJ, as I keep telling you. But I want to build a home. To be a homesteader you have to build your home. And we are going to be homesteaders. If Sue and Steve can do it then we can. I think."

"I think you and JJ can build your own home, CindyLou, if you want to. But you don't have to in order to be a homesteader. It is a great adventure, physically, mentally and psychologically. And for many folks I highly recommend doing so. But you have to look at it practically too. Everyone's needs, desires, resources, and skills are different. There are many ways of being involved in your own home without actually building it yourself."

"And you have to do some thinking and planning first, CindyLou. Sue said *designing* before she said building."

"I heard her, JJ, I am listening. But I want to get moved in before the next century begins."

"Oh, it probably won't take you that long, CindyLou."

"Probably?!"

"Building or remodeling or rebuilding a home is usually a long term project, CindyLou, and it's a fun one. Getting a roof over your head is only the beginning of the adventure. And JJ is right, designing is one of the first steps. Things go a lot better when you have an idea where you're going, and there is a lot involved before hammer and saw begin. But it is an active, involved part of the process so you *will* be doing something. Drawing plans, looking for ideas, talking to people, building small scale models. After all, you don't want your building, whether cabin or house, outhouse or sauna, to collapse upon feeling the first snowfall or strong wind. There is a wealth of information out there to help you succeed. But you do have to take the time to get involved in the learning.

Let your imagination go! Learn, draw, design, play. Make your dreams fit the safety considerations. And make the safety rules fit your dreams. You can let the measuring tape and square and level lie if you want, as long as you don't forsake good old common sense. Dreams and common sense can be great partners you know. Chicken coops can be Taj Mahals, outhouses can be castles, saunas can invoke the invigorating inspiration of simplicity, outbuildings can defy description to those used to established labels."

"Well, that is all well and good, and quite interesting in its own way. But I want to build a home. A real home. Not a cabin or a shop or a sauna. A home. The old shed that is on our land is not good enough to be remodeled into a home so we will need to start from scratch. And I want to start now. You must have started somewhere. Where did you start? Apart from the dreaming and philosophy part that is. I am sure that JJ will take care of that end of it for us. That is assuming we ever get to it."

"I will for sure take care of that part, CindyLou. But we both have to get into the thinking and planning part. It may take awhile. It's a lot of work you know. Maybe we should just consider the sauna first . . . OK, OK, unruffle. A home it will be. But that pile of kindling that you romantically call a shed is not going to contribute much. I can't see using it to build a home. That's going to take materials, CindyLou, lots of materials. And where are we going to get them? And you may not have noticed but there are just two of us, and these old muscles aren't getting any younger, and what'll happen if . . ."

"Sue and Steve are only two people and they build a cabin *and* a house."

"And it took them six years to build the house. So you may as well relax. There's plenty of time."

"Six years?! Did you really build for six years before you had your home? I thought you said you built your cabin in just four weeks? Your home is not *that* much larger. I do enjoy the sawing and hammering but I do not want to be doing that for the next six years!"

"The shell of the cabin did only take four weeks to put up. But we worked on it for many weeks, and years, after that, not to mention the time spent planning it ahead of time. Building is not an afternoon project, CindyLou. Even a simple shelter takes effort and time. How much effort and time depends on what materials you're using, what skills you already have, what you decide to build. It was six years before we moved into the house but we didn't work on it all that time. Nor was it finished when we moved in.

There is a lot to consider before you build anything, or rebuild or remodel, or have someone build something for you. Try to see it as a large kaleidoscope. You turn one way and you see it differently. Change one piece and the picture falls into another pattern. A new idea, a different material, another spot and the colors, the feelings, the patterns take on new dimensions. It is all part of the adventure. The actual building is only one part.

A little time spent on the planning of your project might save the Earth from some negative impact. Or deciding to locate your homestead in a different spot might allow some fellow creature to continue a satisfying life. The nitty gritty is important. But equally so is the philosophical. The spiritual if you will. That's what makes a shell of a building into a home. It doesn't matter if it is a simple tent or a marble castle.

You know, I think the homesteading heroes are, and should be, those who take existing materials and recreate a home that is in harmony with the Earth. Something which is spoken of a great deal, but lived too seldom. Whether it be tearing down an old building, or renovating one that is standing. Rebuilding on an already human-disturbed site, or accepting the challenges of using what has been built before, wherever it is. These are the true homestead warriors."

"Mmmm. She may be right, CindyLou. Guess I'll have to rethink what I think about that spot of yours."

"Another thing to consider is your overall plan. A homestead is an interactive, alive world. You'll want to think of where you'll want your shed and barn and greenhouse and chicken coup."

"And sauna, don't forget the sauna."

"And the sauna, of course. Whatever other buildings, and animals, you plan to be part of your homestead. Garden and orchard too."

"But I do not know what other buildings we are going to want. And I do not know anything about gardens. Or where a chicken would want to live. What if we decide to have a goat? Or a cattle? Oh dear. There is just too much to know. Maybe we should not try to build our own homestead. What if we do it wrong? I do not know if we can do everything that one needs to do. Maybe I am wrong to think I can be a homesteader."

"Of course you can be a homesteader, CindyLou. Homesteads are ever evolving things. They are simply a part of your life. You don't live your entire life all in one day. And you don't build your homestead all in one day either. It is something that is with you and a part of you. It is a philosophy that is reflected in what you choose to do. How you live. Wherever you are. Don't worry about being a particular definition of a homesteader. Live what's in your heart. Just make sure that what is in your heart is what you want to be!

Don't worry about knowing exactly what your homestead is going to be in five, or ten, or fifteen years. Just consider what you *think* you want it to be today. Walk around your spot. Think about what might go where. What *feels* right. It's easy to do. You just have to do it. Let your imagination help you.

At the same time don't worry if down the road you

change your mind. What was built can be unbuilt. There are no set rules here. Each person is different. Each location on Earth is different. Available materials are different. Skills are different. Wants, needs and desires are different.

And you do not have to go into any part of it alone. There is a vast store of information and inspiration out there to draw on. You have to get into it yourself of course, homesteading's not a couch potato sport. But you're already into it, CindyLou. You're thinking, asking, learning. You built your cabin step already, remember? Look and dream ahead, but don't worry about taking more than one step at a time and you'll do just fine. Don't worry if you can't do everything yourself. You don't have to. Being independent doesn't mean being cut off from everyone and everything around you. It means taking responsibility for yourself and your life. And it usually includes a great deal of trading of skills, experiences, knowledge, materials.

You can certainly begin with building your home if that is what you want to do. Maybe if I share with you some of our experiences in planning and building you will get a better feel for doing your own. And realize you two certainly *can* do it. There are many things we would do differently were we to begin today. And no one will want to build exactly what and the way we did. But hopefully we can inspire some ideas for you.

Come on up to the house and I'll get out the photo album. JJ looks like he's ready for a cold drink."

"Sure am. And she's right, CindyLou, you can build whatever you want to build, you'll see."

"*We*, JJ, we. I do think you are right though. We will. Lead on. I am all for action. Or at least photographs of action."

five

Down to Earth, Under the Earth the House

~ ~ ~

"Dreams, plans, drawings, discussions. An idea here, and erasure and redraw there. Another book to read, a different path to research. This is where creating your own home begins. And it's an ambitious, creative, sometimes frustrating, adventure.

We spent hours reading, drawing, building scale models, changing plans. Our energy efficient, passive solar house was going to be great. Slipform stone walls from the Nearings. Underground design ideas from Malcolm Wells. Windows and doors by Eugene Eccli. Basics via the Architectural Graphic Standards. And acres of south facing glass like the best of them. We were ready. Two years and we'd be in. So we thought.

That first unusually hot summer we dug footings and poured the foundation. We appreciated the help from family members on the day of the pour. And we were happy my Dad was there with his 4WD truck to pull the cement truck out our slippery driveway. We knew the cement truck would never come back. I can still feel how good the old muscles felt by the time we were done. We didn't have rock or clay to deal with, thank goodness, but we did have pick and shovel hardpan sand enough to satisfy that part of us that wanted a challenge.

We were very fortunate to have a great building inspector to work with. He viewed our plans with an honest interest, and gave us many good, solid suggestions. He didn't even laugh (at least to us) when he saw our massive footings. Though he did mention that since we were building in well drained sand and the footings would be well buried we could have gotten by with much smaller ones. But they were impressive! At least to those having done the work. And at the end of our first summer that was all we had to be impressed by.

Ah well. Thank goodness, and whichever gods were watching over us, we came up again and again against the barriers of time or money. It ended up being many years before we got to the bulk of the building. By then we had mellowed some, had a much closer feeling for the land, and had come across Mike Oehler's *$50 and Up Underground House* book.

As much as we admired Nearings' buildings, and though we had planned to build the house of stone and cement, we had no natural stone and the price of cement was going up much faster than our house. We lived in the woods, there were sawmills nearby. It was time to bring our focus back to where we were. And Oehler gave us the inspiration and guidelines. Back to the drawings.

The six-sided shape of our place had already been set in concrete (footings). So we adapted our wood design to this shape. It made for quite interesting joints and creative saw work which Steve hadn't dreamed of back in the early days of planning! We came up with a timber-frame structure with 12" x 12" posts and beams, 4" x 6" intermediate posts and 4" x 12" rafters. The ceiling/roof and walls would be two layers of one inch rough-sawn green pine. Because that was what was available, and that was what we could afford.

The original south facing wall plan was a massive 12 foot high by 36 foot wide glass window. Thankfully it never got off the paper. When we stood looking at the site trying to visualize the final building we realized that mass of glass just didn't fit. It didn't matter what the solar design books of the day said. There had to be a better idea.

So we sketched out life-size, right there, our new ideas. Experimenting with various heights using 1 x 4's and 1 x 6's, precariously held together with a nail or two that could be moved in a minute, sometimes on our demand, sometimes on their own, the whole roof outline was sketched out, and the windows. A little higher here, a little lower there. A slightly domed roof/ceiling emerged. And the front window area came down to a livable three foot height.

Now a large south window expanse is great for heating the house on sunny days. The status quo solar home design. But if you actually work and live in your home during winter days it's not practical or comfortable. The glare and heat from all that glass exposure on sunny days would make for a very uncomfortable condition in the south rooms. And that large glass area would allow massive amounts of precious heat to escape during cold winter days. While insulating curtains are a good solution for nighttime, you wouldn't want them closed during the day. But if you understand the dynamics of the situation, and use a good amount of good old moderation and common sense, you can come up with a workable solution. We are happy with ours.

So, what can I share with you from our experiences

with this house? Well, building an underground home has peculiarities of its own to take into consideration. Some obvious, some not. One is that it does need to be insulated. The earth insulates you from the hotter or colder air temperatures (how well depends on the type of soil, how wet or dry it is and how deep). But in colder climates the ambient temperature of the earth can be 45-50 degrees, which is rather cool to most of us for living spaces. So insulation between you and the earth makes for a more comfortable home.

We used an inch of foam board on the bottom half of the walls, two inches on the top half and three inches on the roof. And there is about six inches of soil on the roof. We probably could have used more foam on the roof and walls, but that was what we could afford then. And it works just fine.

[Then again, foam insulation is not an Earth friendly product. Like so many things, there's not an easy answer (or rather, I don't see it, it's probably there). Is it better to use foam insulation thereby using less resources later to heat the place? Or should we work our brains a little more and come up with a better solution? I think the latter. If we build again we will research, think, and design for an insulating material and building design that would be better for *all* of us.]

Never underestimate the power of the Earth. It is truly amazing. Design well for the particular stresses of underground buildings. Then add in a lot of fudge factors. Then overbuild from there. At least that was generally our approach. Believe me, it is quite unnerving to see a 6" x 12" solid wood beam bow in an inch or two. And it takes a *lot* of digging to correct.

Our roof/ceiling turned out to be of a slightly domed design. For looks and to facilitate water runoff. But the regular one inch pine boards we were using didn't want to bend nicely to fit our design. We discovered that a saw kerf half way through the boards worked great over the transition edges of the sharper bends.

Our flooring design was inspired by Oehler's "carpet over plastic" idea. Using a long 2 x 4 float with a long handle we flattened out the sand floor. Back and forth, over and over, back again, forth again, over again, over again. Then another tool, a flat board with a perpendicular handle was made and we started again tamping the sand firm. Back and forth, over and, well, you get the idea. It seemed we tamped forever, but it was not nearly enough as it turned out. Then 10 mil polyethylene plastic was spread from footing to footing (and should have continued on over the footings). Next, styrofoam board was laid out to protect the plastic and provide insulation. Then the subfloor stringers, 2 x 4's on edge on 16" centers, were laid right on the foam.

The we-were-sure-very-flat sand base now turned out to be not so flat. And we played around for some eons with shims leveling the 2 x 4's. Next, we loosened up the old knees for the nailer, and the old arm for the sawyer, and the subfloor grew across the expanse, formerly viewed as somewhat small, now perceived as very large.

At the time all we could get in plain pine tongue and groove was green wood. So green wood we used. The cracks created as the wood dried give the smaller dust bunnies a place to hide, and a spider a ready made fort against an overly inquisitive cat. The creaks that developed as the not-firmly-enough-packed sand base loosened up add interest and music to a quiet home.

At the time we had plans to add hardwood flooring to this pine subfloor. But twelve years later we have come to love and appreciate the pine floor. Some things just turn out to be not that important, given enough time, and other projects.

Wood flooring for the front shop area would have been fine. But we had some used bricks piled up outside. Just the thing. They would soak up the heat from the winter sun pouring through the windows. And they would provide a durable floor. Besides, the bricks had a history. They were from the wall of a local gift shop in town. A new parking lot project turned out one day to provide a bit more excitement than all involved wanted. Workers had gotten a little too close to the wall of the building with their backhoe. Minutes later shelves of glassware and gift items which used to stand against the sturdy brick wall found themselves out in the open air. With the strong brick wall now a chaotic pile of rubble on the ground.

The bricks ended up piled on our homestead as the store owners rebuilt with new. It was time the bricks went back to work. We leveled the sand in the shop area and started hauling them in. Steve placed them firmly in the sand bed and when done swept sand in the cracks between. Instant floor! Well, not too instant. But it did make a nice floor.

However, later it became apparent there was one problem with this design. The bricks were great at wicking up the moisture, and cold, from the ground. No problem with dry winter air with this system! The house plants love it. And it would be great for a greenhouse. But we decided a little less moisture in the house would be preferable.

Out came the bricks just as they went in, one by one. A few inches of sand was dug out, shovel load by shovel load. The bed was leveled. And tamped. Then a layer of polyethylene film was spread over all, then a layer of foam board. The sand was replaced and leveled. And the bricks laid back down. One by one.

It was worth it. The floor has a good live feel to it and is comfortable to work on. The cracks between the bricks can be a problem though, even though they are mostly filled with sand and sawdust. If you drop a small screw or part you can not assume you will find it, things like that love to hide in those cracks. But we like it, and it was (relatively) easy. The cost was right too.

Having the shop area part of the house is convenient. But if we were designing now we would probably go with a separate room. Because we work with wood. And that means sawdust. Everywhere. We're not overly fastidious people, but for practical reasons it would be nice to keep the sawdust in one spot. Which doesn't include on top of everything in the house. But the light, heat, aesthetics, and traffic patterns blend well with the living area of the house so we leave it as is. And dust a lot.

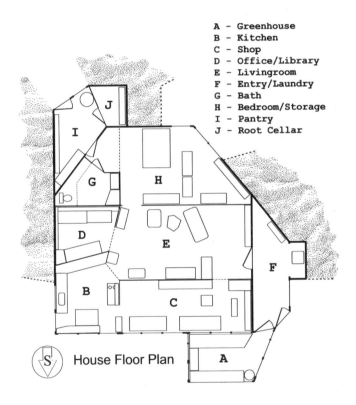

A - Greenhouse
B - Kitchen
C - Shop
D - Office/Library
E - Livingroom
F - Entry/Laundry
G - Bath
H - Bedroom/Storage
I - Pantry
J - Root Cellar

House Floor Plan

I haven't detailed plans of the house but then you won't want to build our house. This is definitely a one of a kind thing. But here is roughly the layout.

I remember when we moved out of the cabin and into the house. I felt like a radish seed who suddenly found herself in a big bean pod. Going from 390 sq ft to 1200 sq ft of space was quite a jump. As time went on our selves, ideas, dreams, and projects filled up the spaces and we grew comfortable here. But two people and two cats don't need this much room to live and work. Next time we would build smaller.

To give an idea of the cost, though not based on current prices, from 1977 to 1992 our expenses for the house were: house - $10,150; water system - $3700; septic/excavation - $700; electric system (solar) - $1840.

We have had a few problems. The biggest we created ourselves by going a shortcut when building. Both the pantry/root cellar roofs and the entryway roof are about an 8" step down from the house. Now Mike Oehler mentions this in his book and says to fill any steps, or irregularities, in the roof area with smooth concrete before putting your plastic down. We didn't want to take the time for concrete so we "filled in" with pieces of foam. And by doing so we made a very nice, cozy raceway for the squirrels and mice. They loved it. Since one of these raceways runs right over our bed we've been aware of the problem for some time. But it wasn't until it started raining *inside* the pantry that we took the time to address the problem.

The nice thing about sandy soil is that is isn't very difficult to dig. The not nice thing about an underground house is that you have to do that digging in order to find, and fix, any roof problems. But digging is good exercise. We found the critters had chewed a substantial hole in our plastic water barrier. And done so in such a way that the drainage from a good portion of the roof was being directed onto the bare roof boards of the pantry. There was a little wood rot but not dangerously so. We patched the spot with generous amounts of tar and plastic and tried to make the area inaccessible and uninviting to the squirrels and mice. And we invited the pine snake who occasionally visits the area to take up residence nearby. As it was fall with winter approaching, and the usual shortage of time, we left it at that for the time being.

We also have a small leak which, we assume, originates around the clerestories, and some problems where the greenhouse roof was attached to the house. So, the big project this summer will be to unbury the whole roof and do some creative redesigning and repair. We do enjoy living this close to the Earth. It fits us and the land. With clerestories and windows it is not at all like "living in a cave" as many imagine. And the buffering effect of the Earth is much appreciated, for temperature extremes and for storms. It is also very quiet which could be a great advantage if one lived in a noisy area. For us we put in one vinyl window to let *in* some of the surrounding sounds.

And then there are the deer and rabbits stomping across the roof at night. Comforting for us, a bit unnerving for first time guests. But it's good to know that the creatures we share this space with don't seem much bothered by our building. We feel that if we were to leave with the doors and windows open, the animals would move in and make very good use of the place. That's a comforting thought.

I encourage you to go ahead and build your own dream house, safely, whatever it is, and with as little negative impact on the Earth as you can. Your considerations will be different than ours. Some things will be more important to you than others; available materials, where you live, energy use, economics, building codes and inspectors, use, water, waste, aesthetics, longevity. There are many aspects to the circle of building, or rebuilding, your own home. The important thing is to jump in and get involved. Consider all the pieces. Then do it."

~ ~ ~ *six* ~ ~ ~

The Wind in the Mill and the Dance of the Water

~ ~ ~

"Here you go. These books should help you out."

"That is an overwhelmingly large stack of books, Sue."

"This is just what we need, CindyLou. As I've been telling you, you have to plan first. You can't just go grab a hammer and saw and start building. Not unless you want to end up with a mouse house. Not that I want to be selfish or anything but I think just the two of us living together's enough. Now, we'll just find a nice comfortable spot and burrow into this glorious pile of information and get started into some serious thinking here. You need to spend more time thinking, CindyLou."

"I *have* been thinking, JJ. And we already have a great amount of information. I do not know that we need a great deal more. Though I have to admit that was quite a long process in your building of your house, Sue. Now I am not sure if we should start with the roof or the floor. And I am not sure I want a pine floor. Not to say that your floor is not a very nice floor. It is just that I had planned to use the white shag carpeting which we had taken up from our old house. It is important to reuse and recycle you know."

"I agree, CindyLou. And if, um, white shag carpeting is what you really want then you should use it. It is your house, you can have it be whatever you want it to be. Though you might want to think ahead a bit on that carpeting as far as maintenance goes. Just from a practical standpoint. I'm sure it's very nice carpeting."

"I was thinking that if we put the carpeting down early on it would be very nice underfoot for working on the rest of the house. Maybe we will build the floor first. Yes, that is what we should do. We will use the boards from the shed to build a floor. Leave the books for later, JJ, let us go build a floor. I would like to pick up the carpeting this weekend. And on the way we can stop by that place that has the tractor for sale. We have to be thinking about our garden you know. And a fence. I have heard that the deer and rabbits here can jump a ten foot fence. We had better get started putting up our fence, JJ."

"Relax, CindyLou, relax. You're getting way ahead of yourself again. I'm sweating all the water out of me just listening to you. And water doesn't grow on trees you know. And this body's not getting any younger either, it needs all its water. Dehydration is a serious problem, the consequences of which . . ."

"Which reminds me, would you like a glass of water, JJ? CindyLou? Fresh from the ground. Via the pump and the storage tank. But refreshing just the same."

"Sounds good to me."

"Thank you, Sue, that would be very nice. Your water comes out of a faucet? I did not know that. I thought you carried your water down from your windmill. We have been carrying our water in jugs from the neighbor down the road. Just look at that, JJ. You turn the faucet and the water comes right out. Just like in the city. And hot water, too. Oops, guess not. That one does not work. That would be very nice to have your water run right into your house. I had forgotten about the water. Maybe that is where we should start, JJ. Instead of a floor we will build a water system today. Not that our neighbors are not very nice people. But it is rather a bother to have to haul all of your water in jugs. Especially when JJ knocks over the last jugfull just when you very much want a drink late at night. And the jug just happens to fall onto *your* side of the bed."

"Now CindyLou, it was an accident. And I'm sure Sue isn't interested in hearing about it. Let's just find out about how they get their water in here. I agree that's more important than a floor. I'm all for not hauling it by hand. These hands aren't getting any younger you know, and those jugs of water will stretch your arms out like a pine snake. Which is not going to be a pretty site in some coming year."

"All you do is lift the jugs into and out of the car, JJ. I do not think you have to worry about your arms stretching any more with that. But I *am* interested in your water system, Sue. I also am not inclined to favor hauling water from a distance long term. But I do hope this is not going to be another long process."

"How long a process getting your water is depends on your resources; where your water supply is, how you choose to tap into it and how you plan to get it into your house. I agree it's important though. Especially before you make a commitment for any land. We've hauled enough water by hand to appreciate the running water. But hardly any water supply is going to be a five minute job, CindyLou. Even if you have a creek or spring to haul water from you'd probably want to build a screened catchment of some sort. Do you know what water sources you have on your land?"

"Well, no, we haven't thought about it really, but there must be water down there in the ground somewhere."

"We do not have a river or a spring. But maybe we can get our water as you do. I have seen the windmill but I do not know exactly what it does. And I had not thought about how it got there."

"It was quite an adventure, CindyLou, and it is something you and JJ may want to consider. Let me see . . . here they are. These are pictures of our windmill project. Most are of the tower going up which was quite an exciting day for us."

"Good heavens! That looks very dangerous to me. I do not think I want to have to do that for water no matter how important it is. Oh my. Look at that giant thing hanging there in the air. Maybe we will forget the water and build the floor instead."

"Like so many things, CindyLou, it's only dangerous if you don't take the time to do it right. That includes planning and learning ahead of time, and careful attention to the details always. People have been putting windmill towers up for a long time and I doubt that very many died doing it. The pictures make it look harder than it actually was. Though I'll admit the process was pretty nerve wracking. But it was fun, too. Why don't we walk up to the windmill and I'll explain our system. It won't be so intimidating once you understand the particulars."

"Looks to me like something that we ought to research a *whole* lot first. I just don't much care for the thought of CindyLou building us something like that, considering the outcome of some of her previous projects."

"*Us*, JJ, us. This is a joint adventure remember. And just where would you be sleeping at night if I had not built us a tent? And the stakes to hold it back down after that first night?"

"Probably in a nice warm comfortable safe bed in a nice warm comfortable safe house back in town. Oh OK, unruffle, unruffle. I like the tent. And we want to have water. Lead on, Sue. I'm listening. Slow down, CindyLou, slow down. That big tinker-toy getup isn't going to fly away before you get there. I sure hope you have some books about all this. It looks to me like it's going to be plenty complicated. I don't know why just getting some water to drink should be hard."

"We have a few books and articles you can read, JJ. But really, your water supply doesn't have to be very complicated. Many people tap into a creek, river, lake or spring. Others get by just fine with collected rain water. And you can always melt snow, though from experience I can tell you that's not as easy as it sounds. And some places you can dig or drive your own well.

In our case we didn't have any surface water to use, other than the small swamp back behind the house, and that wasn't quite what we had in mind. Most wells in our area are drilled. But there are a few driven wells and flows around so we bought a few lengths of pipe and a drill point. Steve made a pounder out of a length of larger pipe and a cap, we rigged a tripod of some sturdy dead trees with a pulley and rope to help lift the pounder, and were ready to drive our well.

Out we went with high expectations and hope. We tried here, we pounded there. We dug down in one depression by the old homestead and put our all into it. The picture of the bent drill point tells the tale. Our land consists of varying levels of sand and sandy-loam over limestone. As it turns out, very deep and very hard limestone. We finally gave up and called the well driller.

Since a windmill needs wind and our house is in the woods, which does a good job of blocking that wind, we chose a spot about 100 feet from the woods out in the field up on a small rise. It took two tries for the well driller to hit water but when he was done we had a 105 ft deep 6" well.

The pump, a beautiful dark red Monitor hand pump, was installed by our well driller along with the drop pipe with brass cylinder and check valve assembly. We had a choice of 3/8" galvanized pipe or 7/16" solid rod for the drop pipe. The price was about the same but we decided on the smaller solid rod because of weight. We would be able to lift that by hand whereas the heavier galvanized pipe would have necessitated using a come-a-long for assistance. In the years since, and more pullings than we like to think about, the solid rod choice was much appreciated. However, a better choice would have been solid stainless steel rod. Which is what we ended up with sixteen years later. But what a joy it was to pump out those first gushes of water by hand.

As fun as that was we still wanted a windmill to help with the bulk of the water lifting. So off we went on our windmill adventure. We had set aside $2000 for the well. And when the well and pump were paid for we had $550 left. We drove the 300 miles back to our city home, called up the Heller-Aller Factory in Napoleon, Ohio to order our windmill, called in work to take another day off, and jumped back in the car.

When we walked into the Heller-Aller Factory we walked into a time much more in tune with where we were headed than where the building was located in the busy city life world outside its walls. Well worn wooden floors, friendly people, windmill parts and pieces organized in an efficient but comfortable manner throughout the several storied old wooden building, still manufacturing and building windmills as they had been for almost 100 years. I'm sure there have been many changes over those years, but the atmosphere was one of a romanticized pre-industrial era. We wandered around soaking it all in.

When we left we had with us our eight foot Baker "Runs in Oil" windmill, and $630 less in the checkbook. A darned good exchange we figured. The half assembled windmill fan was a great highlight of our living room decor all that winter (well, we liked it).

For a water storage tank we purchased a new 1300 gallon round concrete tank, usually used for a septic tank, and had it delivered to a cleared, flattened spot next to where the tower would be. It took some work to patch the spots in the tank where the reinforcing mesh was showing through, and to close the holes in the side. Steve drilled appropriate holes in the bottom for our water line. We found out later that if we had let the person providing the tank know it was going to be for water storage he would have picked out the best tank for that purpose. Instead we got to patch. And it has worked fine ever since.

A friend of mine had an abandoned 60 ft three legged power line tower in their city back yard, a liability they wanted to get rid of. A deal was made, another adventure, and we had the tower in pieces waiting in our own back yard. Thankfully, our city neighbors were quite tolerant of our varied homestead activities and collections.

The tower, along with all the rest of our accumulated things and stuff, moved with us in the spring to the new homestead. One of the first projects on the list was for Steve to rebuild the tower to fit our needs. That and learn to live and work among the sometimes overwhelming hoards of black flies and mosquitoes. With wrench and torch he went to work.

Meanwhile, we marked off and with friends dug three large cone shaped holes (small at the top, large at the bottom) for the tower footings. Angle iron pieces from the original tower were braced in the holes along with cross pieces of reinforcing rod to await the final lifting of the tower. The holes would be filled with concrete after the tower was up and bolted to the angle iron in the holes.

Steve built a platform at the top of the rebuilt tower to stand on when servicing the windmill, and we started painting. We only got the top half painted before it was time to raise the tower (it seemed as if *everything* was taking longer than we expected). But we figured we'd just paint the bottom half later (hah!). We used damp proof red primer which was great for longevity, poor for drying time. But the fellow we'd hired to help us raise the tower was ready. We decided the 'tacky' stage was dry enough.

The "bottom" two legs of the tower were set up on poles and attached firmly to the even more firmly braced angle iron and post stakes in their respective holes. We wanted the bottom to stay right where it was as the rest of the tower was raised, not to scoot across the ground, and into the well pump.

The day came. Bernie and his bulldozer were ready.

Using the front loader of the bulldozer he raised the top of the tower and set it down (gently, gently) on the raised box of his dump truck. A gin pole was rigged, chains and guy ropes attached. The bulldozer was in place to pull the tower up. A rope going in the opposite direction connected the tower to our Land Cruiser (to prevent the tower, once up, from continuing on over). Guy ropes were secured in the other two directions. Bernie manned the bulldozer, Steve the Land Cruiser, I the camera. Held breath ... Racing pulses ... Moving machines and tower ...

      ...

        ...

          ...

            ...

              ...

Whomp. It landed. Right on target. With no surprises. Ahhhhhhhhhhhh. It was fairly quick but that's still a long time to hold your breath.

    Well, there was one surprise. Or something we

hadn't thought of. Someone had to climb the tower to detach the pulling chain and guy ropes. No time to worry about fear of heights. Steve made his first, of many, trips up the narrow tower steps.

The concrete footings and pad were poured June 24, 1978. What would a piece of concrete be without a date or name scratched in it? A great accomplishment, and we admired it fully. Steve saw all his planning and visions and work come to be. I was awed, just beginning to learn about visualization and perspective and seeing a completed project emerge from a bunch of pieces and drawings.

Of course, we weren't exactly done. The tower was up but the windmill itself was still on the ground. Steve built and erected a temporary crane at the top of the tower with a double block and tackle. Some sturdy 1/2" nylon rope, the Land Cruiser, and we were ready to pull the heavy mill gear box to the top. That went well and Steve once again found new muscles as he wrestled with the head fifty-five feet in the air on the small platform (it looked a lot larger when it was on the ground).

We were ready for the fun part. We bolted the fan together and got ready to raise it. Ropes were secured and Steve climbed to the top. I steadied the eight foot fan down below. I realize now that we were blessed with an exceptionally clear, warm summer that year so we were seldom delayed by rain. But now we encountered a different problem. Wind, in the form of a quite gentle breeze.

We had chosen the site for the breeze and wind potential, and the fan was designed to turn in the presence of the slightest wind. And it didn't understand "Not now! Not now!". We lowered the fan back down. And waited for another day.

We learned quickly about the wind patterns of our

new homestead. It's generally calm in the morning until around ten o'clock, then again in the late evening. A combination of daylight, dissipated dew, and no breeze was needed. It took three false starts but we finally got her up. Steve attached the fan. Then came the tail. So few words to describe so much work! But the time finally came when the temporary crane was tossed off the tower and the pullout cable was attached. And Steve could come down.

The pump rod still had to be installed so that the windmill, instead of us, could power the water pump. But the beauty and aesthetics of that windmill were all that counted right then.

Steve made a pump rod out of 3/8" steel reinforcing rod, mainly because we had it. The top of the pump rod connecting to the windmill was made of wood. That way if the pump jammed while the windmill was on (pumping up and down) the wood would break instead of the steel rod jamming up into the heart of the windmill and breaking something there. As it turned out the two pieces of reinforcing rod are joined together through loops made in the ends. So when the pump does jamb (leathers jamming or freezing at the pump) the rod just bends off to the side at the joint. Yet when all is well it pumps up and down just fine."

"Well, now, that's quite a marvel I do admit. But you do get water out of it now, right? Can't say as I'd mind tasting a bit of that liquid myself right now. I got thirsty just hearing about how much work it was putting this thing up."

"Never mind the work or the water, JJ. What is of most importance is if one has to actually climb up to the very top of this very tall edifice in order to make it work? That is a very long way up."

"Well, CindyLou, yes, someone does have to climb to the top occasionally. But not very often. The windmill needs periodic maintenance, such as greasing or oiling the gears. With our mill here Steve climbs up about once a year to change the oil and take care of any other needed maintenance."

"Oh yes, you did mention "runs in oil". I still say it is a terribly long way up, or rather down. Does he by chance wear a blindfold?"

"I certainly hope not, CindyLou! It really isn't that bad once you are used to it. And he does wear a safety belt hooked to the tower when he is climbing out over the edge of the platform, and when he's working up there. If you take reasonable care there's no reason for it to be dangerous."

"Have you ever been up there?"

"Um, well, no. Since Steve goes there hasn't been any reason for me to. I've been halfway up though, does that count?"

"I am not sure. But then, I guess it would not have to be me to go to the top would it?"

"Now don't you go looking at me, CindyLou. I believe in staying firmly attached to the old earth firma here. But about the water. How to you get it out? Do you just turn that faucet there? Just curious you know."

"Oh, sorry, JJ. We'll get you a drink. You can either attach the pump handle like this, and pump by hand. Or, since there is a pretty good breeze right now, we can let the windmill pump the water for us. Disconnect the regular pump handle, then let the pullout cable loose by lifting the long handle over there by the tower leg. Look up at the mill and you can see the tail turn out as I lift the handle.

The tail makes the fan turn into the wind which causes the fan to turn, and the pump rod to lift up and down. Then, assuming there is enough wind, the water will be pumped or lifted out of the well by the cylinder attached at the very bottom. Give it a few minutes to get the water up into the pump . . . Now normally we want the water to be pumped into our storage tank. But when we want to have water here at the pump we just turn the faucet on the side here and . . . Here it is and there you go, JJ."

"Whoa, slower, slower. I just want a drink not a bath! Ah, that's good. Thank you water gods. Nothing is quite as refreshing on a hot day as a nice long drink of fresh cold well water. Here, have a swallow, CindyLou."

"Thank you. That does taste nice. Definitely cold, too. But I have not seen a pump with a faucet on it before. They usually have a spout."

"Yes, that's how they come. In the parks you'll find a spout where the water comes out when you pump it. But we needed to have a way to direct the water into the tank and not out onto the ground whenever the mill was running. At the same time we wanted to be able to get water right here if we wanted. So Steve removed the spout and bolted on a floor flange, then screwed in a regular faucet. On the other side a 1 1/4" flexible water pipe is attached to carry water into the tank."

"What's this other pipe for? The one coming out of the ground next to the pump. Looks like it had a problem of sorts with that split in its side."

"That was an early mistake. Before we poured the cement for the footings and deck (and before the tank was buried) we connected this pipe from the pump to the tank. It goes down underground, up along the outside of the tank, then in through a hole in the upper edge. And it worked just fine for a time. When the pump was pumping the water would be pushed through the pipe and into the tank. Just as we'd planned. Then fall came and freezing weather. Since the pipe goes down before it heads up the water never drains out. And since this end is above ground it froze up the very first good freeze we had, and split the pipe open. Not what we had planned.

We cut it off, installed a new pipe which, as you can see, goes up in an arc and sweeps down into the tank. This way the water in the pipe drains into the tank and/or back into the well. So no frozen pipe."

"But what do you do when it is very cold outside? You can not pump water through your pipe then can you? I have been told to never pack away my long johns because there is not such a thing as summer in the north. That it is cold most of the year. I do not know if I brought along enough layers of warm clothing. And I am sure I will be thirsty during the long winter."

"Oh, CindyLou, it's not that bad. We have great summers. Some are warmer and some are colder but they are still summers to us. There's not snow on the ground right now is there?"

"Well, no."

"See, that's how you know it'll soon be summer. That and the mosquitoes and blackflies will be here soon to officially usher it in. But seriously now . . ."

"I *am* serious."

"You'll get used to the weather, CindyLou, you'll see. But back to the water. It's true we can't pump in very cold weather. The water would freeze right at the pump. So during the winter we only turn the windmill on when it's around 30° and a good breeze is blowing. It depends on the weather. If it is pumping well and the sun in shining then we can have it going when the temperatures are in the mid 20's. Since we can hold 1200 gallons of water in the tank, and we are quite conservative in our water use, that has worked out just fine. We may get low but there is always a break in the weather here and there throughout the winter to allow us to pump some fresh water in. During the warm months we turn the windmill out more or less regularly to keep the tank full."

"What happens if you don't turn the mill off and the tank overflows? That ever happen?"

"Only once, JJ. And we found out what this old pipe was good for! When the tank was full it started pushing water back out of this pipe. It was quite an impressive mini backwards waterfall! But normally we keep a better eye on it. We have a rather straightforward system to measure the water level in the tank. Come on up to the top of the mound here and I'll show you."

"You would have quite a hard time of it if you ever had to unbury your water tank with all of these brambles proliferating all over. Ouch, darn. Does this water tank (I will take your word for it that it is indeed under here) ever freeze? How would you get your water out then? Oops, ouch again! Give me a hand up, JJ, this grass is slippery."

"Here you go, CindyLou. You shouldn't grab those brambles without gloves on you know. Interesting little building you have here, Sue. Who lives in it?"

"No one in particular I hope, JJ. But one time we had a very strong wind and found the roof blown off and lying on the ground. It looked like a small earth bermed house. We scrambled down to get it, joking about who could live in it. Just as we lifted it up, a slight warning aroma hit our senses. We dropped the roof and backed off quickly, but not before a resident skunk got the message of his displeasure across quite emphatically. Luckily for us, he only hit Steve's boots, though it was quite awhile before those boots were allowed into the house. We decided we would retrieve the roof later—which we did with no further incident.

It's actually quite a heavy little roof built with one inch pine, so blowing off isn't often a problem—that was the only time. It slides on over the walls like a lid. This building protects the opening, and cover, into the top of the tank. If we need to get into the tank, we lift off the roof and open the trap door, which is large enough to allow you to put a ladder down. It's not something we have to do often, but every once in awhile we drain the tank and get in and clean it out well. We poured a cement lip around the opening into the tank, inside this box, to prevent dirt and critters from washing into the tank. The roof keeps it dry from above.

But we didn't want to have to get in and open up the large lid just to measure how much water we have in the tank so we came up with this system. This coffee can is caulked into a hole in the roof and if you look down in there . . . It's a little dark but you can see a small metal pipe coming up almost to the roof. That goes through a hole in the tank lid and gives us a place to stick our measuring stick. And when done we just put the plastic lid back on the can so nothing else can get in.

To measure we take this stick, first look to make sure it doesn't have any unwanted hitchhikers, and put it down through the coffee can into the pipe and down into the tank. The cord at the top keeps you from losing the stick in the tank since it goes down below the roof. Pull it out and there you have it, see, marked on the side. You can see we have about 850 gallons in the tank. So we'll leave the windmill out and let it pump for awhile."

"But how did you know where to put your marks to say how much water you have? It seems a little, if you will excuse the expression, crude."

"I don't mind the expression at all, CindyLou, and I think I know what you mean. You could have a more complicated and expensive system, but why? This works just fine. We just used math formulas based on the size of our tank to figure out at which height different amounts of water would be. Then we transferred those marks to our stick with a file and with a permanent marking pen, every 100 gallons. Simple and it works. After awhile you get a feel for how much water will be pumped with a certain amount of wind, and how much you're using, so you don't have to measure all that often."

"Well, I did not think I would have to learn math in order to have my homestead. It seems that everything which is simple is in fact very much complicated. I am not very good at math and neither is JJ."

"That's what friends are for, CindyLou. And books,

and libraries. I didn't know how to calculate the volume of water in our tank either but I learned. That's half the fun of this whole adventure, the learning."

"She's right, CindyLou. Next time we're in the library we can look it up and I'll show you. I can't remember the formula exactly but I know where to look. Why with a few math formulas you can figure the volume of all sorts of things. How much air is in our tent, how much peanut butter is left in the jar, how much steam it'll take to fill our sauna, how much . . ."

"I am not going to spend all day in a library just to learn how much peanut butter is left in the jar. I have eyes. All I want to know is how do we figure out how big a tank we will need for our water system. Why is it that everything has to take so much time? I will never get my homestead at this rate."

"Don't worry, CindyLou, some things sound more complicated that they really are when you get down to doing them. Ready to go down?"

"Yep. Whoa there, watch yourself, CindyLou, it's a mite slippery coming down."

"Move aside, JJ, here I come. Yippeeeeeeeee! That is not a bad small ski slope you have there, Sue."

"Thanks. One of these days we'll get some steps built."

"Now the water just flows on down to the house from here? No pumps or anything? Doesn't seem like that would work."

"No pumps, just gravity. You use what you have and in this case we had enough of a drop from the tank to the house for the water to flow. You really don't need much. We have a line buried down into the garden and get a good enough flow there even. And, as you can see, it's not a large drop. Though at the top of the garden you don't get a strong flow out of the hose unless the tank is full. We don't water the garden that much so it's not a big problem"

"What are those pipes for? The ones sticking up out

of this side of the tank hill?"

"At the very bottom of the tank Steve installed good large gate valves. One for the line to the house and one for the line to the garden. He then built a small box around the valves with 4" PVC pipe coming out of the top. A long handled custom made valve turner can be put down the pipe in order to turn the valves on or off so we can shut the water off to either the house or the garden. Theoretically. It's a great idea. But when the tank was buried the dirt pushed the pipes so hard against the valves that they can't be turned. So we have no way to turn the water off up here."

"So what happens if you have a leak or have to fix something?"

"Well, you can drain the tank, which takes a long time and an even longer time to fill. So we don't do that often. What actually happens is you work like the devil when you have to cut into a water line or pull apart a connection and try to make that very cold, wet, spitting-turning-to-gushing shower time as short as possible. It's the other person laughing so hard at the sight that probably causes the worst problem. And every time we swear we are going to dig out those valves and redo the design so we *can* turn the water off up here. Just haven't gotten to it yet."

"Seems to be an awful lot that you don't get to. Do you ever get anything completely done? Seems to me that with the proper amount of research and reading ahead of time everything ought to run just hunky dory."

"Oh things run pretty good in spite of the way it seems. Sometimes you get the hunky right and sometime later you usually get the dory. And in the best of conditions you get them together at the same time! Or at least not too many years apart."

"But does not anyone ever just build their homestead without all of these, well, side tracks and interrupting difficulties? I am thinking that I may never get my homestead adventure underway and now it is looking as if there is no way to ever get to the end of it!" "Don't despair, CindyLou. It's the path that's the adventure, not the final product. Besides, when you're talking about homesteading I don't think there is such a thing as a finished product. Just think of it as a bunch of mini-adventures. Each side track and interruption and difficulty is a part of the whole. While it's important to keep your eye on the whole picture and where you want to go, you only have to live one moment at a time, one adventure at a time. More or less. The important thing is to take each piece a bit at a time and enjoy doing whatever you are doing. It really is fun, trust me."

"Seems to me the one thing you keep coming up with every time is work. And more work. And more sweat. Speaking of which, how about we lounge ourselves down over under that apple tree?"

"Really JJ, there is not a thing wrong with work. There is not a part of your body which could not benefit from some good exercise."

"I have to agree with CindyLou, JJ. The part about there being nothing wrong with work. I'm not insinuating you need exercise. But accomplishing something with your own hands and your own muscles and your own mind and your own spirit is something which is truly a great experience."

"I'm not saying work isn't good for a body, or so they say, I'm not overly convinced myself, but I'm just rather wondering who it is we're going to get to be doing all this work. OK, OK, don't glare at me CindyLou. It's hard on the skin. So, just to be thinking of where we might be heading, what else should we know about your water system. What other, uh, adventures have you had with it?"

"The biggest problem we've had is because of our hard water. Great for taste but hard on equipment. With our original steel drop rod we had a big problem with mineral scale building up on it, and it was being pitted and eaten away at a rather alarming rate. Then the particles would fall down and get packed in the leathers. This caused them to push out and jamb against the inside wall of the brass cylinder at the bottom of the drop pipe. Which stopped the whole operation cold. So we would have to pull the leathers out. And with 85 feet of rod this wasn't a two minute job. Though not bad in the warm months, I can't say as it was my favorite activity in the winter. But we did get very good at it, sometimes having to do the job a half dozen times in a year. We tried a number of combinations of leathers, one, two, half, whole. But it still jammed. We settled on using just one leather which worked fine most of the time.

We hadn't heard of this problem from others, which seemed strange considering the number of windmills that used to be in operation. But in talking with a neighbor farmer we came up with a possible reason. He said that when he was a youngster they had a windmill to pump water for the cattle and he didn't remember having to do much of anything with it. Certainly not pull the leathers to clean them. But the mill was in operation most of the time because of the cattle's needs. We figure that probably kept the particles in the water riled up and moving so they didn't have much of a chance to settle down and wedge in the leathers. And with a large family, common when windmills were common, the house mill would also be pumping a lot. Since we pump sometimes only two or three times a month the particles in our water have a good long time to settle and pack into our leathers.

But we were getting tired of pulling the leathers. And our pump rod was getting dangerously thin in places. Time for a different approach. One option was to replace the steel rod with fiberglass rod but the possibility of having fiberglass shreds in our water didn't thrill us. Instead we checked around, dug deep into the old pockets, and installed new 3/8" stainless steel rod, in 12 foot sections. A friend threaded the ends for us.

The stainless steel is heavier. It is about all we can do to lift and drop it by hand. But it is shiny and smooth. And the extra cost for stainless steel well worth it. It should last a lot longer than the regular steel rod did.

Beautiful as it is though, it didn't solve our leather jamming problem. A great disappointment. We didn't have the flaking from the steel rod so it wasn't as bad. But we still had the very fine sediment from the water. What now? One more idea. We pulled the rod once again. Steve took the pump leather and pared off most of it, leaving just the bottom ring with a small lip. Ever hopeful we dropped the pipe back down, put the pump back together, let the tail out, and prayed. And we must have connected with the right gods this time because the mill pumped and the water gushed out as usual. That little bit of leather was enough. And there was no place for sediment to get into to jamb. It worked! We haven't had to pull the leathers since. A skill which we are quite happy to set aside.

There are not a whole lot of moving parts in the windmill, but they do all have their jobs to do. And, as with the leathers, it doesn't take a large piece to fail to bring the pumping to a halt. And that seems to always happen in the winter. One December I went up to the pumping windmill to turn it off to find the pullout cable had worn through and was draped over the pullout handle. Well, at least it was a different problem. There was no way to pull the tail out and shut off the windmill. Not a good thing. We unhooked the pump rod from the pump and tied it off with an elastic cord to keep some tension on it (a loose flying pump rod can cause all sorts of unwanted damage up in the mill). Then we waited for a not too windy, not too snowy, not too icy, not too cold day for Steve to climb to the top and reattach the cable.

A good day finally came. Steve couldn't find his climbing strap. Another wait while we purchased new climbing strapping and clips. Another day, up he went, repairs were made. Homestead joys always do come, they are just sometimes delayed.

Another piece we replaced with stainless steel was the short rod that connects the drop rod to the pump rod through a brass nut in the top of the pump. The original steel rod soon rusted and pitted and was tearing up the pump packing (which ended up down in the leathers to add to that problem). And due to a not quite straight welding job by the well driller (we all learned on this job) the pitted rod rubbed against the brass nut and wore it out. We replaced the nut, $12, and replaced the steel rod with a nice shiny stainless steel model, $70. But twelve years later it is still shiny and smooth, the nut is in good shape, and our pump packing lasts longer."

"I am thinking that hauling water from the neighbors is not such a bad thing. Not that I do not think your windmill is a very picturesque addition to your homestead, Sue, but it does seem to take a great deal of effort for just a simple drink of water."

"Well now, CindyLou, how long would you want to go without a good drink of water? The neighbors may not want us to be bugging them for water forever. And you get a mite cranky when you don't have water to wash up with. Course, the way things are going we may not have to worry about good water. To drink or for anything else. Won't matter how much work you want to get into to get it if there's none to be got. Why just the other day I read about . . ."

"Take it easy, JJ. Don't give up on the world yet, OK? I agree the overall picture can get pretty gloomy. But if you look at what each of us *can* do and what we *are* doing then you can find quite a bit of hope. Look around you. Not much to complain about here is there?"

"True, true. But water doesn't grow on trees you know."

"Well, in a sense it does, JJ. But back to the practical aspect of water for your homestead. It did take a bit of effort to put our water system in. And it takes some more now and then to keep it running. But most of the time it runs just fine without us thinking much at all about it. Or doing much other than to turn the windmill out when we need to pump more water. And turn the faucet on when we want it.

But this isn't the only solution for pumping water. Instead of a windmill you could have a solar operated pumping system, which would be much closer to ground level and take little maintenance once installed. It would pump whenever the sun shone and the water level was down. And you could arrange your piping to be all underground so you wouldn't have to worry about freeze-up when pumping in cold weather.

Another option in the non-freezing months is rainwater. Catching the rain from your roofs is pretty straight forward and simple. You could come up with quite a nice system with a little ingenuity. And not much money. For many people rain water *is* their source of water. We even utilize it, in a very simple manner, even though we have the other system."

"That is what all those buckets are for lined up behind your shop building? I did wonder but I did not want to pry."

"Seems we're kind of beyond worrying about prying, CindyLou, what with all our questions and all."

"That's OK, JJ, I don't mind your questions. I'm glad to be able to help you in your adventure. But, yes, CindyLou, that is what all the buckets and that large garbage can is for - to collect rainwater running off the roof of our workshop. Since our well water is quite hard I like the soft rainwater for washing clothes, and us. It's great for the hair.

In the winter you can melt snow, and we've done that. One year our well went dry, the water dropped below the level of our cylinder. We hauled water from town and melted snow all winter. One of those adventures I don't care to repeat. It takes a *lot* of snow

71

to melt down into much water, and it is quite dirty. We were sure glad that year when spring came and we could get down with the pipe to put an extension onto our well, and pump out of the ground once again. But the melted snow worked and we would do it again if needed.

So you can get by without drilling a well, using rain water or snow or carrying from another source. But a working well near your place is a very nice convenience."

"Yes, I think you are right. And we will have to think about it. And, even though I hate to admit it, do some research. But we could even now figure out a way to catch rainwater off of our tent roof. That would be very nice to have soft water to wash with. Some of my clothes are not very happy with the hard water in this area. Yes, that is what we will do. This afternoon. Let us go, JJ. Let us go set up our very own very first water system on the CindyLou and JJ Homestead. There does not seem to be a lack of rain here. And we could get some pretty colored buckets to set along our tent. Yes, I can just see it. Pink and blue and green and yellow. This is very exciting. Let us go, JJ."

"Well, I guess it would save us some hauling. But I don't know about having pink buckets outside the tent. Couldn't you just go with those old white ones we got from the neighbors? They're good enough for Sue and Steve. Pink. Why do you have to have pink? Not a very natural color you know, and what if some bird or skunk or something gets offended? You never know . . ."

"Why don't you take some water along with you? You're welcome to haul water from the well here anytime you want. I'll go get a couple of buckets."

"Thank you, that would be very nice. We are very happy to have you offer. This is so exciting, JJ. We, JJ and CindyLou, are going to have our very own homestead water system. And good water from your well, Sue, thank you again. Fill your bucket there, JJ, and let us get going. The day is waning."

"Now hold on there, CindyLou, you don't have to be

in such a darned hurry. My bucket's not full yet. And besides, all that hurrying is going to cause you problems some day you know. And this cold water splashing down my leg will probably cause me arthritis. Not healthy, not healthy at all. Would you slow down. I'm not getting any younger you know, and these legs . . ."

"Take care JJ and CindyLou. Come back again soon."

"*Ohhhhh, we'll have us a watersystem, the rain can begiiiinnnnn, for drinks wash and rinsing, and cleaning your chiiinnnnn . . .*"

~ ~ ~ *seven* ~ ~ ~

Touching Toes With Insects
The Get Down and Get Dirty New Garden

~ ~ ~

"*Mmm hum mmm mm humm. Hm mmm hm mm hm mm hhhmm.* Oh! Hi there JJ and CindyLou, didn't see you coming."

"Hello Sue. We are sorry to interrupt your . . . work. Or your humming. Are you by chance singing, or humming rather, to your plants? I have heard of doing that. But I have never seen, or heard, anyone actually, that is out loud . . . That was not bad humming."

"You should talk, CindyLou. You're always singing out loud for no reason at all, enough to wake every sleeping mosquito between here and Canada. As if there weren't enough hoards awake already. At least Sue was just humming."

"I did not say there was anything *wrong* with it, JJ. I was just wondering. And besides, Sue says she enjoys my singing. And the mosquito situation is not my fault."

"Yes, CindyLou, you heard me in a rare act of humming out loud to the coming plants. It just seemed like a great day for it. If I could sing as well as you do I'd sing out loud for them. But generally I stick to just talking with them in my head. They hear me just the same. So what brings you two out on such a gorgeous day?"

"Oh, CindyLou has it in her head that we ought to be having a garden. Myself I think we should to stick to real food like canned beans and spaghettios. Stuff we know. Leave all this digging and scratching to folks who *like* to get down on their hands and knees and get dirty. Never know who you're going to run into down there in that dirt you know."

"JJ, if we are going to take care of ourselves and be responsible for ourselves we have to be able to feed ourselves. We have talked about this before. If we are going to be proper homesteaders we have to be sufficient unto ourselves for our food. All of the books say so. And Sue seems to get a great deal of pleasure out of it. So we are going to make our garden this afternoon. We just came over to find out all we need to know to grow our own food. We won't bother you for long."

"Glad to help. Digging down into the dirt of the garden with your hands is a great way to meet all sorts of new and interesting creatures. And yes, I do enjoy it a great deal. Gardening is something that just gets into your blood. You'll see."

"I think I would rather not get quite so much of the garden on me let alone in me. But I suppose one could wear gloves?"

"Oh, this dirt washes off easily enough. But of course you can wear gloves. Whatever works for you. I'll be happy to share what I can to help you get started on your gardening adventure. But you know, there are tons of books written about gardening and growing food, and they only begin to touch the subject. I think it will take you more than a few hours with me and an afternoon on your plot to learn how to grow your food."

"See, I told you, CindyLou. We have to start at the library. You can't just go digging in the ground and expect to get food. I say we think about it this year and maybe next . . ."

"I am not going to spend all summer researching how to grow food. People have been feeding themselves with their own hands for eons without one library at their disposal. It cannot be that hard. I am going to have a garden *this* summer."

"You're both right. It isn't that hard once you have a basic idea of what you are doing. And yes, people have been putting food on their plates for centuries using their own hands and feet. People all over the world are doing it right now. But in the past, and in some cultures today, the knowledge of food growing and gathering was common among the whole community. Children grew up amidst this rich world of experience knowing how to provide food for themselves and their families. Unfortunately, not many of us have that opportunity today. So we learn where we can. I've done it and so can you. It takes some effort but it's worth it."

"You keep saying that and it always ends up being a whole lot of work."

"Which is well worth it you have to admit, JJ. You two are well along on your plans for a more permanent shelter, you have your own unique if temporary water system and you certainly can get started on your garden. It just isn't going to happen all at once, CindyLou. Our garden here has been evolving for almost twenty years. You just find a place to start and then dig in, searching, learning, experimenting as you go."

"But I do not want an experiment, I want food. And I do want to make our garden this afternoon. Before the air gets any more crowded with biting flying inhabitants. It would be all right if our garden looked like yours, for now. Though maybe we could make it just a bit neater without all those bumps and, well, you do not seem to have much dirt showing. You did say this was your garden?"

"Sure is. And there's no reason you can't get started making your own this afternoon. I think that once you get into it you'll get an idea of what all is involved. And, JJ, once you've eaten your own homegrown beans and potatoes and tomatoes and corn those factory produced cans and boxes of food are not going to taste much like real food to you anymore. You've survived quite well eating at our house haven't you? And nearly all of that food came out of this piece of ground which, in spite of CindyLou's doubts, really is my garden."

"I guess I have to admit that you've served us some mighty fine eating. Just never thought of where that stuff came from. But it's too nice a day to be thinking about digging and all. Maybe we should wait until fall when it's cooler and these giant attack bugs are full and ready for a truce. Back, back! Do they ever give up?! I don't have much blood left already. I think we ought to crawl under a blanket back in the tent and contemplate life."

"If we contemplated life as much as you wish to, JJ, we would never live life at all. And even I know you can not plant your garden in the fall. You have to do it right now. So we will. And the mosquitoes and black flies are not that bad. Just ignore them."

"Ignore them?! You kidding? They'd suck me dry before you could sing two notes of a song! Just 'cause I taste better than you doesn't mean you need to get uppity about it. Besides, these critters could be carrying all kinds of disease you know. Malaria is not something to be taken lightly."

"I think you don't have to worry much about malaria up here, JJ. And, CindyLou's right, if you can take your mind off them they aren't that bad. You'll get used to them. But meantime we have some headnets I'll lend you which may help. And I agree contemplating life is important. As long as you don't forget to live it, too! Maybe the black flies and mosquitoes are here to remind us that we're living. Anyway, you can start a garden most anytime the ground is dry enough. And not frozen of course. And now's as good a time as any. You'll probably be able to get some crops in yet this year. It's just not all going to happen in one day.

How about if I give you a tour of the garden here and give you some ideas on how to get started. I'll even explain what the bumps are all about, and why there's so little dirt showing. And introduce you to some of the resident spiders and insects."

"*Spiders and Insects!?* What do you mean resident spiders and insects? You don't invite them here on purpose do you? Watch your step, CindyLou. If they grow the garden bugs as large as they grow their mosquitoes and black flies we're in big trouble."

"*Our* mosquitoes and black flies, JJ, ours. If we are going to be living here we have to accept everyone and everything else, too. Though I must say I had not actually thought about spiders and insects being in my garden. There are not too many I do hope."

"Well, if you have a healthy garden there will be a healthy population of insects and spiders, too. This garden is full of them, nice critters they are too. Some cause me grief to be sure but they are a minority and I deal with them one on one as they come."

"That sounds a mite more intimate than I care to be with anything I don't know, and I don't plan to get to know them any better. Maybe we better get back to work on the cabin, CindyLou, and leave your garden for winter when Sue's friends are all frozen."

"*Our* garden, JJ, our garden. You are always telling me about some creature or another that we have caused to be extinct from the Earth. Well, here is a chance to get to know some who are left. Though I must say I am not of a mind to want to actually touch them. Not that I wish to be rude but holding hands with the insects was not exactly what I had in mind when I decided to have a garden."

"The insects are pretty much involved in living their lives and I doubt they are particularly interested in becoming close friends of ours. If you just let them live their lives and you live yours, and exchange pleasantries when you meet, you'll get along just fine. They are nothing to get upset about, they are just a part of the garden world. You'll get used to them, you'll see. Just like the black flies and mosquitoes and wood ticks."

"Wood ticks? What's a wood tick?"

"Oh, don't worry about them yet, JJ. You'll meet them soon enough."

"We will not worry about the insects, JJ. They will not kill you. Well, not usually. But I do wish to get on to the subject of building us a garden. I hope this is not going to be as long as the subject of building a shelter was. Not that I am not very grateful for your help, Sue. I am very happy that we are working on the plans for our cabin. But I do want to build our garden and grow our food this afternoon."

"It doesn't have to be as long a discussion as building a home was. But I'll warn you, gardening is an almost infinite subject. Probably more so than any other homesteading activity. Just get a few gardeners together and you'll see what I mean. It's more open ended than building a house. There are numerous beginnings and an unending number of endings. And a whole lot of adventure in between.

And you're right, you don't have to worry about the insects. You'll no doubt read and hear all sorts of horror stories about them and gardens. But the only ones you have to deal with are the ones you'll meet in your own garden. And most of those are friendly. The others you take one by one. There is really no need to be scared or violent where the insects are concerned.

I can't tell you how to grow all your food this

afternoon. But I can help you get started. There's no reason you can't grow something to eat this year. Though I wouldn't depend on it for your entire food supply just yet. But if you work at it, each year's garden will be better than the previous. And before you know it you *will* be providing all your own food with plenty to share. Keep that in your mind, but don't belittle what you are able to grow the first years. It'll be the best lettuce and beans you've every tasted, believe me!

Here, have a garden sorrel leaf to nibble on. It's a great perennial and I'll dig you up a chunk of it to plant in your new garden when you're ready. It's one of the earliest, and latest, crops in my garden. And all you have to do is let it grow. Let's sit down here in the shade and I'll see what I can tell you. I won't get into the particulars of growing each crop right now. There are a great number of books and articles on that subject and I'm sure JJ will be digging into those once you get into this adventure. Besides, they'll be easier to talk about later when the plants are up and growing.

One thing is for sure, there is no one right way to garden. I'll share with you how I started the various additions to this garden and you can start the same way if you want to. You'll soon be experimenting with various techniques on your own to find out what works best for you. But it's nice to have an idea of where to start. When I first gardened I bought two gardening books and that was all I needed. I've added a few more to my library over the years but this isn't something which requires a great deal of money or research. Talking with other gardeners will probably be your biggest source for information and ideas.

The first thing to do is to choose your garden spot. Now all the books will tell you the ideals, and also tell you to save your most fertile ground for almost every crop they list! Trouble is, we have to deal with what we have. And it usually doesn't resemble the theoretical ideal, and is seldom very fertile. Few gardeners have the formula ideal site, yet they are all growing good food and a lot of it. But you do want to make best use of what you have of course. No use making unnecessary problems for yourself and your garden by putting it in the worst spot.

In our area we have to deal more with wet conditions than dry so a spot with good air circulation is important. And good sun most of the day. Some shade is nice for some crops but for the majority, sun, and its heat, is a plus. Our biggest challenge most years is to get the heat loving plants to maturity before the first frosts. And, while air circulation is good, it's not a bad idea to have some break from the winds from the north and either the east or west.

Now I'd read that having your garden on a south facing slope of ground was good, the cold air drains down away into the lowest part. We did that here. And we found that it does indeed work for the light frosts and light cold, *if* your garden is at the *top* of the slope. Our first garden was below where it is now and sure enough, the top rows would escape damage while the lower ones would get frosted even on this gentle rise. So we moved the beds from the face of the rise more to the top of it, and I keep the most tender crops in the upper areas.

It is written that you shouldn't plant near to trees because either the tree will take all the moisture or the plants will take all the moisture leaving none for the other. Well, I think the trees and plants know how to live together better than that. I've found that under this apple tree is a great place to grow greens. The tree shades them from the heat of the sun in summer and they coexist just fine.

As you can see there are several bushes and fruit trees within the garden. We hadn't planned it that way but the garden grew up around them as it expanded. I plant up quite close to them all. Less ground to have to mulch that way. The bushes are healthy and so are the plants around them. In fact, the first year in that area there, which was a fairly dry one and the soil was not very fertile yet, corn near to the Pea Shrub and the Lilac Bush did better than those plants farther away. Now if you're going to be using a rototiller you'll not want to till up close to or very deep around the trees and shrubs. And you want to leave enough room to harvest both fruit and vegetables.

Another important consideration for us is fencing. If you want to plant your garden for the deer and rabbits and raccoons then that is fine. But if you want to harvest most for yourselves you need to have a good fence. We put our fence up before the garden went in to protect our newly planted apple trees. And we managed to pick a fairly decent spot. When we started gardening it was within this fence and we have never had a problem with deer in the garden (except when I've left the gate open which they appreciate and take full advantage of).

The wire is a standard six foot woven wire fence with smaller openings at the bottom (to keep smaller animals out or in). The posts are cedar. When we put it up we bent the bottom 6" out flat on the ground to keep critters from digging under the fence. That leaves us with a 5 ½ foot high fence. The deer could jump it if they wanted to. But since they have never gotten used to eating out of our garden they don't bother I guess. Many gardeners in deer country who have put fences up after the fact have had a heck of a time keeping the deer out, ending up with very tall fences.

Now fencing will not do much against ground hogs since they come in under ground. We had one dig his tunnel to come up right in the middle of one of the best green bean patches I've ever grown. His last meal was a feast to be sure, and we had few green beans for ourselves that year. The only solution I know of is ground hog soup.

Raccoons are particularly frustrating neighbors. We had no trouble with them for the first ten years or so. Then the forest land around us was traded to a paper company by the Forest Service. They came in that summer with their machines, chipped up all the trees and hauled them away by the semi truck load (that's one reason all the books we publish are printed on 100% recycled paper). It was a sad summer for us, but especially so for all of the forest inhabitants who used to live there. Soon we had raccoons in the garden, in the compost pile, at the bird feeder. But mostly, in the corn.

After much loss of corn and many trials we finally came up with a solution in the form of a short, five-wire electric fence on pole posts with regular electric fence standoffs. We put it just around the corn when it's starting to ripen. The strands are 4", 8", 14", 22" and 30" from the ground. The middle one is grounded so if the coon does climb the fence he will still get zapped. In order to work the animal has to have one part grounded and the other touching the electrified wire.

We found that it takes a fairly high voltage to get through the coon's thick fur coat. More of a setting for a sheep than a cow. We also discovered that they, at least our coons, must check it out nightly. One night with a corn stalk or weed fallen across the wires, shorting out the fence, and they are in there. Feasting, partying and no doubt having a grand time.

Before the electric fence I tried the oft recommended squash around the corn. The theory being that the coons won't cross the prickly squash vines. It didn't work with our crew. We would come out in the morning to find piles of corn cobs and coon scats right in the middle of a web of squash vines. I think they were being sarcastic. That doesn't mean you shouldn't try it. One thing I've found in talking with other gardeners is that there is definitely no one rule for all. What does not work in one garden may work fine in another. What one swears by is a disaster for the next. You just have to choose the techniques that interest you and give them a go in your own garden.

We also tried live trapping. But it was soon apparent

that we had quite a healthy herd of the beasts. And when we, apparently, captured Momma coon and she tore up our trap we gave up on that approach. As with the ground hog, raccoon roast could be a solution, if you don't mind eating it every night for quite awhile (holding to the maxim that if you kill it you eat it). For us the electric fence has proven to be a good solution.

You can build your own fencer yourself if you want to. Steve did for our first one. See *Home Power* issue #21 for a schematic. But we had trouble getting it set right. When it died we bought a regular electric fencer from the farm supply store, hooked it up with an old car battery and a solar panel to charge it, and set it up by the fence for the duration of the corn harvest.

Another consideration in choosing your garden spot is water. If you plan to water your garden you'll want to think about where that water will come from, and how you'll get it to your plot. Over the years I've come to watering less and less, finding that the garden does just fine without my interference. It adapts quite well to what moisture Nature provides, even if it doesn't come in the formula one inch per week. I might water when I plant if it's particularly dry. Or when the plants are just emerging. Seldom after. However, most of the garden is kept well mulched which makes a lot of difference in how much water it needs.

Will you be hauling your water from a creek or lake or cistern? From rain runoff or waste water from the house? From your well? And how can you get the water from the source to the garden plot? In our case we ran a line underground from the water tank to a faucet in the garden where a hose can be attached. There isn't much water pressure at the top plots since the level isn't much lower than that of the tank. But it's enough. Last year we finally installed a hydrant so we don't have to worry about freeze-up at the faucet in the winter. It works great. A lot better than burying the faucet in hay bales every fall.

And then there is the ground itself. Uh, JJ? CindyLou? You two awake there?"

"Of course I am awake, Sue. I have listened carefully to all you have said. But I am getting discouraged. It does not sound as if I will have my garden this afternoon."

"You have to research this stuff, CindyLou. Think it out. And what a great day for thinking. Just look at those swallows swooping around up there. Now if it weren't for these whining little blood suckers it'd be just grand. Hold on there! You're taking more than your share! Look at that. Boy, does it itch. A body's blood doesn't grow on trees you know! Away, away! Back off you buggers, back off!"

"Hey, watch what it is you are swatting, JJ!"

"Hold on a minute, JJ . . . Here, try some of this. It's an herbal based insect repellent. It doesn't last as long as the other stuff but I think it's safer. It seems to help. Or at least it makes us think it helps which is close enough. You do get used to the mosquitoes. But I'm not sure one ever gets used to black flies. They are more a matter of endurance.

Think of it this way. If there weren't any black flies or mosquitoes the swallows wouldn't come. Or the bluebirds. In years past we've had more than a dozen swallows swooping the air. That's why we have so many bird houses around the garden. But the mosquito and black fly population has been dropping and now we have only a few birds coming. I hope this is just the natural dip in a natural cycle. While I can't say as I exactly miss having more of the biting insects I do miss the birds. Let's head over and look at your future garden site. Being on the move will help some."

"What a choice. Chompers and birds or no chompers and no birds. Well, this stuff smells like it ought to do something! Want some, CindyLou?"

"No thank you. I think I will put up with the bites instead. And I do not care to think about having to deal with more of these particular insects. Though I do like the birds. Look, there is a bluebird! It is a bluebird is it not?"

"Yes, we have one pair that comes every year. And though they often do raise a brood we seldom see more than the one pair. They sure brighten up a day! I hadn't seen a bluebird since I was a youngster, when DDT use was common, until we moved up here. I appreciate that they decided to take up residence in our garden. They arrive about four days before the swallows do and are here after the swallows have moved on. Nice neighbors to have."

"That is another good reason to have a fence. To have a place to hang the birdhouses. Yes, we will have to get our fence up, JJ, so we can put up some swallow and bluebird houses. Boards from the old shed would be just right for that. We will make us some beautiful bird houses for the beautiful birds."

"I suppose we'll be doing that, CindyLou, I suppose we will. Anything having to do with sawing and hammering seems to draw you like a mosquito to a warm body. What I want to know is how come we come over here to get a question answered and always go away with more questions and projects and work? Our conversations always end up with you wanting to build something. This body's not getting any younger you know. I don't know how many of these sessions it can handle! If we aren't careful we're going to end up . . ."

"Being healthy and happy right up to the day you die!"

"Well, that wasn't exactly what I was going to say, Sue."

"But if you continue with this homesteading that is just what might happen, JJ."

"I agree. But we must get moving or the sun is going to go down without you and I even getting a start on our garden, JJ. And we will not have any food to eat this winter."

"Oh, don't worry about that, CindyLou. We usually have enough in our garden to share. Since you and JJ don't have a plot worked up yet how about if you take over a bed in our garden? Then you can plant whatever you want and get a start on learning to garden, and growing food. Then you can work on getting your own plot ready for next year. It really works much better if you take at least a year to prepare the soil before you start planting. Making a good garden is more than an afternoon project, CindyLou."

"I was afraid it was going to come to that. We do not want to be a bother to you and Steve. But that would be very nice to be able to plant some seeds this year so we can eat. Thank you. And we will of course help you in your garden in return. Right, JJ? JJ!"

"Well, there you go adding more work already and we haven't even gotten to our place yet. I don't want to be disneighborly or anything but this summer is getting shorter by the minute and it's barely started."

"Don't worry about helping us right now. You two have quite enough lined up already. Our garden is in pretty good shape and you'll be busy enough just getting yours started. And planting and tending your one bed. Is this the spot?"

"It is indeed. Can you not just picture the rows of muskmelons and watermelons and sweet potatoes and tomatoes . . ."

"Of course she can't, CindyLou, give her a break. All that's here is weeds! *Lots* of weeds. And sod. I can't see that this will ever grow a pumpkin let along a year's worth of food. Of course, we'll probably just drop over from all the work of trying to get to the dirt that may or may not be underneath all of this grass and brambles and trees and brush. I say we should just go get a goat."

"Well, let's just take one thing at a time. This isn't a bad spot you've picked out here, CindyLou. But those small cherry and poplar trees aren't going to be easy to get out by hand. Not that you couldn't do it. But have you considered any other spots?"

"Yes, I suppose the trees might be in the way. But you have trees in your garden, Sue, and you said they were not a problem. I think these trees are cute and would look very good in my garden."

"But what I have are fruit trees and shrubs, CindyLou. These small trees are going to grow into large trees. It won't be long and they will be taking up more room and casting more shade than you will want in your garden. Having a tree within your garden can be very nice. But you have quite a few of them here. A more open spot might be better if you have the option."

"There is another area which I have looked at. Maybe it would be better. It is over here and it is not far from where our cabin will be. As soon as JJ stops drawing plans so we can get to the building."

"This is a great choice, CindyLou. You have some woods to the north and west which will block some of the strong winds. Yet far enough away to give the breezes room to play. And the grass looks healthy, too. I can picture a very nice garden here."

"You might as well get on with it. I know you're going to get to the digging and weeding and all that work sooner or later."

"Of course, JJ. That's part of the fun. There are a number of ways to take a spot of natural field and turn it into a garden plot. We are fortunate in that we're working with a sandy-loam soil. Folks who have to carve a garden out of boulders and clay have a much tougher time of it. But don't worry. This will be challenge enough."

"I wasn't exactly worried about it being enough of a challenge."

"Well, I wouldn't want you to get bored with it all, JJ. One way to clear a spot is with a garden fork, and a good back. This is actually one of the better ways of taking care of the natural vegetation, or what we term 'weeds'. Start in one corner, loosen a small square of sod, knock off as much of the dirt as you can and toss the weeds aside in a pile, or into a wheelbarrow or basket to be hauled off to a pile. Once these are composted they'll go back on the garden. You just keep on going down one side then on across your plot. So much easier said than done of course! When you have all the vegetation off go back across the plot loosening up the dirt with your fork.

I've done that for a small area and it works well. But for a larger garden it is rather time consuming and quite a bit of hard work. Another way is to mulch the sod with hay or straw, pieces of carpet, old (natural fiber) clothing or blankets, cardboard, paper bags, whatever organic matter you might have. Whatever you use it has to go on very thick to smother a good healthy sod, and it will take at least a season, maybe more. But the next year it will be easier to dig. This can work very well if you have the materials, and the time to let them work. Be sure, however, that whatever you put on as mulch will either compost without leaving any harmful residue, or will be easily removed without leaving any non-compostable pieces, such as zippers in old jeans, plastic linings in feed bags, tape and labels on cardboard boxes. There will be different options for mulch materials in different areas. Hay is readily available here most years.

Then there is mechanical digging and this is how we've worked up most of our garden areas. With a rototiller. Since you can choose to care for your garden by hand after the initial breaking up of the sod you could rent or borrow a tiller for this purpose. You don't have to buy one. You two can borrow ours if you want. You may want to use it several times throughout the first summer. The initial tilling up, and in, of the sod is just the start. With that operation you break up many roots into pieces which will grow again. And you're bringing to the surface many weed seed which have just been waiting for this chance to grow. You will want to till these back into the soil when they start growing. This subsequent tillage could also be done by hand.

However you break up the initial sod it will be good to work in several green manure crops throughout the season. Not only for fertility and tilth but also to knock back the weed growth. Green manure is simply growing plant material worked into the soil. Weeds make good green manure. If you haven't anything to plant simply let the natural vegetation grow up again and turn it under before it goes to seed, then do it again. This is a method which is probably as old as agriculture. And it works as well today as it did in days past. But if you can you'll want to plant a crop to turn in to get more plant matter into the soil.

What to plant depends on what is available. You'll be planting a large area, not just a row, so inexpensive is (usually) important. I plant what is available from the local farmers; field peas, oats, buckwheat, soybeans. I prefer oats over rye because it winterkills and I don't have to worry about it growing up next spring. Buckwheat is nice because it'll grow a good crop in poor soil, and the bees love the blossoms if you let it grow that long. I stay away from clover because it takes a longer time to grow a good crop, and it is very hard to get rid off if you are working your plot by hand. However, if you have several years and will be working it in by machine it could be a good choice. Vetch is another option. Local farmers and feed mills are a good place to go for recommendations.

Ideally you won't let your green manure crop get too tall before digging or tilling it in, especially if you are doing it by hand with a shovel. Maybe knee high or so. However, if it does by chance get taller, which happens more often than not to me, you can mow it first with either a scythe or a mechanical mower. Then till or turn it in. Or leave it lay and just plant your next green manure crop right into the trash. This works quite well too. The residue isn't incorporated into the soil but it will still decompose over time and provide tilth and nutrients.

Probably one of the more important chores is to convince yourself that a plot of ground full of or covered with plant trash is a healthy looking spot, whereas a garden bare of plant material is anemic. Nature seldom leaves the ground bare and she knows more about gardening than we can even begin to guess.

It is good to plan your plantings so you will have the ground covered over winter also. Either with something growing in or plant material covering the soil. I often plant oats in August so it will be 4"- 10" tall going into winter. It will winterkill and in the spring I loosen that ground and plant right into the oat trash. I've found that if I let it get much taller it mulches the soil too well in the spring and keeps it from warming up. Then when I rake it off to plant the roots pull out, which I would rather have in the ground. So if it gets too tall I scythe it down in the fall and leave it as mulch.

For your plot here you could rototill it up now and plant field peas and oats. You can get the seed from our neighbor. When it is lush but not too tall you can till that in and plant buckwheat, it should be warm enough by then (buckwheat is a warm weather crop and will be killed by frost). You'll have to check around to see if anyone has any buckwheat seed, or maybe I have some left you can use. The farmers don't grow it very often around here. And you should be able to get in another green manure crop after that, using whatever seed you have or can get.

And don't forget the humble but essential compost pile. That is something you can build right now, it doesn't have to be fancy. And every time you add something you will know you are adding to your future garden. We use a simple three bin system for our compost. We add kitchen scraps and garden leftovers to the first bin for a year, then add to the second bin the second year, the third bin the third. After a year or so the first bin is composted and ready for use. When that bin is empty of compost we start over filling it up.

We don't turn or dig or fuss with the compost, we just let it set. It does just fine on its own. Half of the compost is used inside the greenhouse for potting soil, the other half in the garden, dug in when transplanting tomatoes, peppers, and squash, and on top of the seed potatoes. Since I leave most of the garden refuse in the garden I don't have a great deal of compost, or a great deal of work.

You can also get to work putting up a fence, and planning your plot. Where do you want to put your strawberries, and rhubarb, and herbs? Do you want any fruit trees or bushes? One thing's for sure, you'll probably change it many times over the years, so you may want to keep that flexibility in mind. Just don't forget those fresh from the ground sweet crunchy carrots, and sun warmed juicy strawberries, and first taste of spring rhubarb, and can't be had from the store fresh full flavored green corn, and . . ."

"Enough, enough! I'm drooling all over already! You got me, don't overdo it. Guess we'll be getting at your garden right soon, CindyLou. Where did you put that shovel?"

"*Our* garden, JJ, our garden. Thank you, Sue. I believe I can see it happening. I do not see why they call this the simple life though. It is not seeming very simple to me."

"I often wonder that myself, CindyLou. But in a larger sense it is simple. Nature provides what we really need quite generously. That just doesn't mean we don't have to work for it. But the work is much of the fun. You just have to get over the idea that work is drudgery. It's all in how you look at it."

"Well, now, I think it's all in the sweat and dirt and blisters myself. But then again, I can just taste those fresh cooked sweeeeet sweet potatoes. And crisp coooooooool muskmelons and . . . Come on CindyLou, let's get a move on. We're not getting any younger you know."

"Say, I think we should talk a bit about the growing season here, and what will and won't grow. And . . ."

"Maybe we can talk about that later, Sue. We would love to sit and talk with you but we do have a great many chores to do. And the sun does not stay up all night even for us homesteaders. Now, JJ, we must start with the fence. I do not want the deer and rabbits eating our orange trees."

"Oranges? Ah, CindyLou? JJ? Oh well, I'll be talking with you again. See you later."

eight

The Joy of (Solar) Cooking

~ ~ ~

"Greetings, Sue. Do you need some assistance in carrying that, um, whatever it is you are carrying?"

"Hi CindyLou, JJ. Thanks but I can get it. It's unwieldy but not heavy. If you would like to turn that stand a bit towards me, that's good, thanks. There."

"There? There what? A funny box with a glass door? I hate to ask, because every time I ask CindyLou decides to have whatever it is and I end up with another something to build. A body's only got a certain number of projects in him you know. And hours and minutes don't grow on trees and . . ."

"We all have the same number of hours and minutes, JJ. And where would we be had I not asked questions? We would still be eating beans and spaghettios out of a can. And we would not have our own water supply. And we would not have begun our cabin. And . . ."

"OK, OK, I get your point. You can unruffle."

"Sue has shared with us a very many number of good projects. Although I do admit they always seem to take much more time than it appears they should. However, I too am wondering, what is the purpose of that bulky box?"

"This box is a simple way for you to cook your beans without having to fire up that old gas stove you have. This is our solar oven. Or part of it. Just a minute, I'll be right back."

"There. It wouldn't do much without the reflectors. That's it. Aim it towards the sun, put your pot of beans and potatoes or whatever inside. Then come back at dinner time and it's ready."

"Well, now, that doesn't sound too difficult. Maybe this is something we should have, CindyLou."

"I see now what it is. I think. But I do think there probably is more to it than just what you said. There usually is. How does it work?"

"As far as cooking in it there isn't much more, CindyLou. If you are going to be around you can re-aim the oven at the sun throughout the day. If you can't do that then you position it to get the most sun closer to when you'll want your food done then leave it there. And you can throw a blanket or such over the glass after the sun's down to keep the food warm if you will be eating later. You have to experiment with your own oven and the food you want to cook to see how long they take.

And of course you need good sun. Partly cloudy can be OK, it will just take longer. If you have many clouds or a strong haze the oven won't get enough sun to cook well. But on sunny days you put your food in, aim, and wait. What you can cook depends on how efficient your oven is (how hot it gets), the size of the cooking box, and the amount of sun it receives.

It's also great for heating water. Whenever there isn't something cooking in our oven there is usually a jar or pot of water heating. When hot we put it in Thermoses to have hot water available whenever we want it. We use the Thermoses whenever we heat water no matter what the heat source, saves time and energy. The oven is also useful for making up batches of natural dyes, especially with stuff and mordants you don't want to boil up inside the house."

"I think I do not want to know about what you can not cook inside, or why. But I thought you said a solar oven is simple. This does not look simple to me. And if it is so simple why do not many more people use one? I have not seen one before."

"It can be quite simple, CindyLou. Once you understand the various parts of it you can come up with whatever materials you have or can get to make your own. Each homemade solar oven is a little different than every other one. Maybe that is one of the things that makes it so much fun. The idea has been around for a long time and used by many people. But to make your own is still very much a creative process. As to why more people don't have and use one? The answer depends on the society and culture you are talking about. No matter where you are it is hard to break old traditions and habits, no matter how much sense the new idea makes. And it does take a different planning and using than the usual gas or electric range."

"How did you find out about them if they are not common?"

"We have to thank Karen Perez and Kathleen Jarschke-Schultz and the other folks at *Home Power* magazine for introducing Steve and I, and so many other folks, to solar ovens. And their enthusiasm for getting us off our excuses and into the shop to make our own. Once made, the fun of using it comes easily. And the knowledge of how gentle it is to the earth compared to many other cooking methods helps to keep the interest high. *Home Power* has run many articles on solar ovens and held Solar Oven Design Contests which prompted many great designs. It is a continual and ever evolving tool. However, you can buy them ready made too if building your own is not of interest to you."

"Since we don't have much of that stuff called money I guess we'll have to stick to the build your own crowd. And I can tell by the look in CindyLou's eye that we'll sure enough be building one. All you have to do is say "build" and she's interested as a fox in a chicken. You might as well tell us how you do it."

"Be glad to JJ. Our oven's design was dictated by the materials we had around. But many solar ovens are built using just cardboard boxes, aluminum foil, and glue. Plus some odds and ends that help out the finished product. Let me get my pot of beans in the oven here to start cooking, then we can go inside and I'll show you how to make your own.

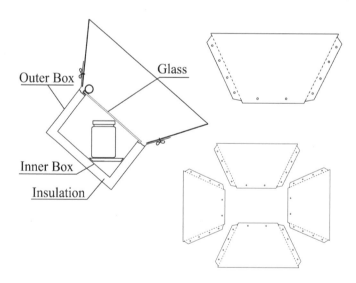

Outer Box
Glass
Inner Box
Insulation

A solar oven can be built many different ways. Once you get the basic idea you can expand and experiment. Use what materials you have on hand or can salvage from around your area and come up with your own design. Start cooking in it, then modify and redesign from there. Meantime, you're cooking a lot of meals with the sun.

There aren't many parts to worry about. Let's look at them one at a time.

Cooking Area/Inner Box - some say it should be black, some say reflective, some don't mention it, I'm not sure it matters. But it does need to be able to handle hot oven temperatures without melting or outgassing into your food. It has to be large enough to hold your cooking pot or pan or jar or whatever when the oven is tipped at the angle necessary to be aimed at the sun. Get a box and a pot out in the sun and experiment, you'll get the idea. But don't get hung up on using a particular pan or pot. A smaller area is easier to heat (within reason) so maybe you'll want to change cooking utensils to fit the size box you have. Then again, you'll want to use what you have so try to fit your oven to that. You may as well assume you will spill, so consider how you will clean out the area when that happens.

Outer Oven Box - I use the word box loosely but I'm referring to whatever it is that you come up with to "house" your oven. The outside part. The part that will be exposed to the weather. It can be anything from a cardboard box to the ground to a washtub to an old packing crate. It needs to (1) hold the cooking area box, (2) contain the insulating material, (3) hold up the lid/glass/top, (4) be able to hold up to whatever use, abuse or weather it will be exposed to.

That doesn't mean it has to be rain or snowproof if you plan to move your oven in out of the weather when not in use or during bad weather. A full time outdoor oven will need to be weatherproof. Your basic cardboard and aluminum foil cooker is a fair weather only affair. If you're going to use your oven much you will probably want to come up with one you can leave outside. As you saw, they tend to be rather unwieldy beasts to be hauling around much. Though you can build or buy smallish, compact models too.

Window/Door - In most oven designs the oven glass is also the door. However, your design could make the glass fixed and have the door be in the back, or side. Often the oven is built around whatever is available for the glass, whether it be an old car window or glass from a discarded window pane. It does need to be able to hold up to the opening and closing required to use the oven. Some type of knob or handle to assist in opening is nice, glued on or hooked on one edge. Remember, this glass is going to be hot when in use.

Top Frame - This is as varied as the designs. But you may need a frame to hook/set your glass on, and attach the reflectors to. It will also cover and protect the insulating material which is between the two boxes.

Insulation - Let your imagination loose. Hay, other dried vegetation, newspaper, blanket, cardboard, fiberglass duct board, wool, whatever. Can it handle the warmth? Is it nontoxic? Will it provide enough insulation in the space available? Can it be kept dry? Is it readily, and inexpensively, available? Try what you have. If it doesn't work either redesign your oven to accommodate it or try something else.

Reflectors - Often the most difficult part of the oven is the reflectors. But the success of your oven depends a great deal on how they perform. It can be as simple as one flat rectangular reflector up to complicated parabolic designs. Within reason, the more and better the reflector area the higher temperatures you will achieve in your oven. So most designs have reflectors surrounding the window. The angle wanted for most efficient sun reflection depends on where you live and what time of year it is. A 60° to 70° angle works in our area.

The size (height) of your reflectors will also be a factor in the angle needed. If your reflectors don't seem to reflect a good amount of sun into your oven change the angle or make them larger. Do use care where you look when adjusting and fine tuning your reflectors, and when using the oven. You want to cook your food not your retinas.

Practical material for reflectors is limited by the requirements - it has to be shiny and be able to be cut to shape. By far and away the most common is aluminum foil glued to something sturdier, such as cardboard, wood, or metal. For our oven we originally bought what looked to be quite shiny aluminum, wanting something more durable than aluminum foil, but it proved to be not so shiny when put to use. We ended up gluing aluminum foil to it. Our new one will use mirror stainless steel.

Some recycled options for material would be opened and flattened aluminum or coated steel cans, such as pop cans or kerosene cans.

You also need to have some way to attach the reflectors to the oven - punched holes and string if a cardboard model, or brackets or a ledge for a sturdier design. If of lightweight material you may need additional ties from reflector to box to keep them from flapping or self destructing.

Cooking Pots - Most anything will do, though a non-reflective container would be better. I've used cast iron, enamel ware, clear glass, dark glass. They all did just fine. If you use a glass jar poke a hole in the metal lid to let steam escape. Quart jars work well for hot water. The problem is more one of size and shape to fit your oven than the material. You may need to make your own cookie/biscuit sheet to get one to fit. Setting it up on something might give you more room.

Inner Shelf - This may be as simple as a few rocks or a board or a shallow pan to a self leveling swinging shelf. Or maybe your design doesn't require one. Most, however, will need something to set your pots on so they sit level even when the oven itself is propped at an angle to get the best sun exposure.

Miscellaneous - Lightweight designs such as those made of cardboard will probably need some anchors to keep it from blowing away. Some bricks or rocks inside may be enough along with the food. Or it may need some additional twine tie downs. A nearby hook or shelf, either part of the oven or not, is handy for hot pads and stirring spoon. If you attach an outside shelf to your oven keep in mind that it will need to be adjustable if your oven needs to be tilted to aim at the sun. I assure you, a shelf sitting at 10° does not hold a spoon or pot very well. We need to redesign that part of our oven. Depending on your oven you may also need blocks or shims to adjust the oven angle so it aims directly at the sun.

Our oven was built using what we had around. A large cardboard postal tray is the outer box. An old stainless steel steamer pan that we picked up at a garage sale fits inside with enough room for several layers of fiberglass duct board, which we also had. We have plenty of odds and ends of glass around so Steve cut a piece to fit for the window/door.

The whole thing has gone through (and continues to go through) many modifications and changes but right now it has a custom plywood top, hinged glass lid (using glass door pivot hinges), a wooden knob handle, several coats of old fiberglass resin and oil based paint on the outside cardboard box (which kept moisture out for about a year), hinged metal shelf painted flat black (as is the entire inside), and aluminum foil covered aluminum reflectors on four sides.

The whole thing rests in a wooden base mounted on a pipe pounded into the ground so we can easily track it as the sun moves across the sky. The seasonal angle adjustments are made with a scrap of wood jammed in behind the oven at just the right spot. It cooks many meals for us from spring through fall. When it gets cold enough for us to have a wood fire going regularly in the house we move our cooking indoors to the top of the heating stove. And put the oven away in the shed for the winter. Steve used small bolts and brackets to connect the reflectors one to the other which makes it easy to take apart and store flat.

Our biggest problem is that our outer box is not very waterproof. The whole thing is getting mungy after being out in the rain for so many years, and the damp insulation doesn't insulate as well. So Steve is building a new oven with better reflectors and a sturdier, more waterproof, box. But meantime, this one cooks and heats water quite well.

Now, are you ready to build one for yourselves? We'll head out to the shed to find some suitable cardboard boxes. I'm pretty sure there is some wide aluminum foil under the counter there left over from our oven. We don't use it much for anything else. And here are scissors, knife, glue. And a protractor to help set the angle for our reflectors. Now, to the shed!

Here you go, JJ. This box will do for the inner. It's large enough for a pan or a couple of squat jars but small enough to fit in a larger box. And I think, yes, here is a large sturdy one that will be good for the outer box. Over here are some large pieces that will do for reflectors and insulation. These held photovoltaic panels but you could also check out places that sell large appliances for big pieces of cardboard. There's glass stacked outside here beside the shed. This piece looks like a good size. We can cut it down to fit. OK, back to the house.

Now it's just a matter of cutting and gluing and assembling. Here's a yardstick you can use to measure and cut against. Just slide this board under your cardboard so you don't cut into the floor. See, CindyLou, not *all* homestead projects take weeks to finish. And you have to admit, JJ, this isn't all that much work is it?"

"I will have to take your word for it that all of these pieces will end up being a solar oven that we will be able to cook our food in. And that we will finish yet today."

"Well now, knocking around on these old knees isn't what I'd call just easy. But I guess I have to admit it's not as much work as some of your ideas. Ach, I'm not sure getting up is any better than staying down. Hey, take a care there with that knife, CindyLou! It's not as if I have extra toes to contribute to the project you know!"

"I am sorry, JJ, but I did not come close to your toes. You should not be walking on our future solar oven any way. Now, what do I do with this piece, Sue?"

"First we will cut pieces of cardboard to fit the bottom of the outer box. As many pieces as will bring the inner box up to a half inch or so below the top. Then more pieces cut to fit will fill the space between the two boxes. If we don't have enough cardboard we can use crumpled newspaper. The glass we will cut to fit over the inner box but inside the larger one, with enough space to fit our fingers or a catch to lift it. If we have any suitable black paint we can paint the inside box. I think there might be some in the entryway.

Now for the reflectors. Grab the yardstick and that square and the protractor. The bottoms will match the size of the outer box. We can attach the reflectors to the flaps of the box. If we make them about the same height as the box is wide that should be about right. We can fold any extra out and down to strengthen the tops. Use the protractor to cut out a cardboard template for our 65° angle. OK. Now, help me hold this piece up here, JJ. Set the angle template along the top of the box, CindyLou. And we'll set our piece out to match it. There, how does that look? OK, trust me, it looks great.

We can tie a nut or something to a string for a plumb bob and hold it to the top edge of our reflector. There, this will do. Measure how far out from the box it is. That's how much wider the adjacent side reflectors will be at the tops than the bottoms. Now do the same thing to the shorter side to find how much wider the longer side reflectors are at their tops. Draw that out on the cardboard, adding that additional amount on either side at the top, then add another inch along the sides to attach each reflector to each other. Look OK? Well, I think it looks OK so lets cut them out and hold them up for a trial fit.

Whallha! A solar oven! Well, almost. To make that top fold it helps to crease along the line with something like the closed scissors. We can glue that down. Now, cut and glue the aluminum foil to the inside face of the reflectors. We can bend these inch side tabs out, poke holes in them and lace the sides one to the other. Or we can bend them in and overlap them on the inside, poke a few holes and tie the pieces together. You'll want to be able to easily take them apart for storage. And poke more holes around the bottom of the reflectors and the outer box while you're at it so you can wire or tie the reflectors to the box.

Let's measure for our glass and get that cut to fit. Somewhat traumatic but it does work if you have a good glass cutter. Usually. Set it on the workbench here, hold the yardstick down firmly, one good strong zip. Hold that piece down there, JJ. And, there she breaks! See? Wasn't so hard. We'll smooth off the cut edges, carefully, with some 220 wet/dry sandpaper. The window/door is ready. What do you think now?"

"I think we are ready to cook our dinner! We have built a very nice solar oven have we not, JJ?."

"I can just smell that old bean stew already. My stomach is rumbling. Let's get it going CindyLou."

"Just one more thing. You have to have sun, and enough hours of sun, to cook in your solar oven. Each oven's different and you'll have to experiment to see how long it takes to cook each dish. I think it might be a little late in the day to cook dry beans. But you can set it out in what sun is left and warm up some water. Meantime, we'd be happy to have you share our solar cooked bean dinner. And I'm sure we'll have another sunny day soon so you can get cooking in your own."

"Fine with me, I'm all for not waiting too long for dinner. We did do quite a spot of work building this oven and a person has to keep his strength up."

"I guess we will have to wait to cook a meal, but I

would like to set our oven out in the sun anyway. It does not look right sitting inside here with the sun shining outside. Come along, JJ. I want to see it in front of our almost cabin. We will be back in plenty of time for dinner. You will not starve. Oops! It is larger than I thought. Oh, worry. Here, you carry it, JJ, and I will get the doors.

Thank you, Sue, we will see you later. Be careful now, JJ, this is our oven you are carrying not a pile of firewood. *Ohhhh, a solarovencookeddinner we'll soon have of our owwwwnnnnn, with fresh roasted carrots, and cookies and scoooonnnnes."*

"That's a terrible thing you're doing to that song, CindyLou. And who ever heard of fresh roasted carrots? Fresh roasted steak, now, that's something a man can get . . . Hey, you're supposed to guide me *around* those branches! I can't see around this dang thing. We should've taken the reflectors off. We almost there? A body's only got so many steps in him, you know, and we've been using up an awful lot of them lately. If it's not one thing it's another. And a body's got to eat too. Been a long time since lunch. Ach, drat. Why didn't you tell me I was heading into that bush? CindyLou? Hey, CindyLou? Where'd you go?"

~ ~ ~ *nine* ~ ~ ~

Outhouse, Inhouse

~ ~ ~

"Eeeeekk! Jump for the bushes, JJ!"

"Sorry, CindyLou, I didn't mean to startle you. I didn't see you standing there. Don't worry, I'm not going to run you over."

"Get a hold of yourself, CindyLou. Sue's not nearly as big as a bear would be."

"No, I guess she is not. I am sorry for my outburst. I know you have said there is not much likelihood of meeting a bear in the woods but you did sound like something very large crashing towards me."

"I know, it's OK, CindyLou. It's just this cart load of firewood. I loaded it up a little full and it tends to push me instead of me pulling it when going downhill. And this path is not exactly smooth. I'm sure you'll get comfortable with being in the woods without worrying about bears before long. I've only ever seen tracks and we've been here twenty years. But what brings you two out this way? Just enjoying this beautiful summer day?"

"Well, it *is* a very nice day to take a walk. But that is not why we are here. I was just . . . that is . . . you see the woods is very good for . . . what I mean is . . ."

"Just say it, CindyLou, just say it! It's not like Sue and Steve have never, you know . . . You do go camping all the time in the woods so you must . . . Well, you don't take your outhouse with you when you go!"

"Oh, I see, I didn't see the trowel in your hand. Well, don't let me interrupt. By all means, continue on into the woods. I'll just take my load of firewood on up to the house. And yes, we do use the woods for our bathroom when away from regular facilities. Works great, better sometimes than the indoor jobs. As long as you bury it well in the top six inches or so of soil, and use leaves or such instead of toilet paper, it doesn't hurt anything. In this sparsely populated and visited forest that is. But please, don't let me keep you."

"Oh, that is OK, we were on our way out of the woods, not going in. I would not have wanted to meet a bear, or thought I was meeting a bear, on the way in!"

"AAAAmen!"

"But to be quite honest this is not the most convenient way to take care of this particular, uh, chore."

"Not to mention the mosquitoes and black flies sucking you dry while you're at it!"

"That is true. And soon we will be buried in snow and then it will be even worse."

"I don't think you have to worry about snow yet, CindyLou. We still have a lot of summer yet and fall too. Thank goodness, because we have a lot to get done before the snow flies. Including filling up our woodshed."

"We also have to do that. Our pile is not very large though I have hauled many poles out of the woods and JJ has cut quite a few dead trees up. But our cabin is not done yet and we have much work to do on that also."

"Now don't get all worried about it, CindyLou. We're going to get it all done long before winter, you'll see. But maybe we should take some time off from cabin building to do some outhouse building. I don't want to be having to head out into the woods in the cold and snow at all hours for certain particular activities, if you know what I mean. A person could shiver to death before they were done."

"I know what you mean, JJ, but the snow doesn't usually arrive until November, and that's a ways off yet. Your blood will thicken up by then anyway. Next winter when it's twenty below you'll think 40° a heat wave."

"Twenty below?! Twenty below what? You do not surely mean twenty below zero degrees on the thermometer do you? I would not survive that."

"Of course you'll survive, CindyLou. And you'll enjoy it too, you'll see. It will be all crisp and beautiful and white sparkley out. But you don't have to worry about that now. Just enjoy the warmth today. By then you'll be snug in your cabin with a nice fire to keep you warm. However, an outhouse would be a nice touch even now. It doesn't have to be fancy. It can be temporary until you have more time to build a better one. Or indoor facilities if you're so inclined. It's a great luxury whether Nature is dropping mosquitoes, rain, or snow on you."

"I do not think my blood will ever be so thick I will enjoy twenty degrees below zero temperatures. But I will wait and see. The thought does make one appreciate especially these warm days. Relatively warm days that is. Down south this would not be considered warm. I would like an outhouse building before it gets deep with snow and before it even starts to get that cold. However, I do not see how we can finish our cabin and get all of our firewood in and work on our garden *and* build another building."

"Boy, I didn't think I'd ever hear you say you *don't* want to build something, CindyLou."

"It is not funny, JJ. Sometimes I think I am not up for this homestead adventuring. Maybe I am not cut out to be a homesteader."

"You don't have to be anyone other than yourself, CindyLou, and that is more than adequate. You are doing just fine. How about coming on down to the house for a strawberry break? They're ripe and warm on the vines just waiting to be picked. That's sure to make you feel better. I'll meet you down there."

"Pick yourself a snack and settle in, I'll get us some water . . . Here you go."

"Thanks. Now this is the way to view summer. Right from the ground, surrounded by food."

"And mosquitoes."

"Yeah. But I think I'm getting used to them. Sort of. Anyway, they aren't as bad as they were earlier."

"You two have come a long way. Think of what you've done. Your cabin is well on its way. You have a very nice comfortable setup. And look at all that is growing in your bed here in our garden! You can be quite proud of what you've accomplished in just these short months. You are already in your homestead adventure, CindyLou, and you are doing very well."

"Yeah, we're not doing so bad for a couple of newbies to the woods homesteaders, CindyLou. I think we've been working pretty darned hard. Maybe we should take a few weeks off and . . ."

"JJ! The snow is coming and our cabin is not done and our firewood pile is very small and we do not have enough food stored up for even a few weeks let alone a very long winter and we do not even know what to do for a building for our outhouse and . . ."

"OK, OK, unruffle. It was just an idea. I know we've got a few things yet to do, but we'll get there."

"I know how you feel, CindyLou. But it isn't as bad as it seems. I remember feeling the same way for many, many years. Until I realized it really didn't matter. Whatever we got done each year was enough. And worrying about it never made it any better or go any faster. You do what you can and make do with what you can't. And it's a good idea to take time off once in awhile. I can see why you wouldn't want to take off for two weeks right now, but it's important to get away when you need to, even if it is only for an afternoon walk. All the building and everything else will wait. It's more important to enjoy what you are doing. And CindyLou, you already *are* a homesteader. It's as much a state of mind as any particular thing that you do."

"Thank you, maybe you are right. I *am* enjoying our adventure, it just sometimes gets rather overwhelming. But I do have to agree that these strawberries are a great boost. We will have to have strawberries in our garden, JJ. A lot of them. But I guess I will have to wait for them. But maybe we should not wait to build our outhouse. Though I do not like to think of taking time out of working on the cabin to do so."

"You know, I think you have a solution at hand. And it will not take you much time."

"Well, no offence and all, but I think I've heard that one before. And it always seems to lead to a lot more work than what it sounds like at the beginning of your projects."

"Now JJ, they have helped us very much with our cabin. And I *would* like to have an outhouse nearby even right now. One like yours, Sue, with a window would be very nice. Though I think it is not necessary to have the entire door of glass. I do not like to think of the mice looking in at my toes. But how can we have an outhouse that will not take much time?"

"A tent. Our first outhouse was an old army tent that Steve had. We used it for many years until it fell apart. It didn't have a glass window but you could leave the door open for a good view. We had to keep the snow shoveled away from it in the winter but it kept the rain and snow off your head. And it was handy for storing many other things besides the, well, the biffypost."

"A tent? That was not quite what I had in mind. You are not thinking of *our* tent are you? And what is biffypost?"

"Now hold on a minute, CindyLou, that's not too bad an idea there. We're going to be out of the tent soon, real soon I hope because it's getting a mite small nowadays with all that stuff you keep hauling home and piling in so soon there's not going to be room for us anyway, and . . ."

"That is not *stuff*, JJ. There are many important homestead items which I have procured without which we will not be able to survive this homestead adventure."

"Uh, CindyLou, I think you're supposed to *enjoy* your adventure not survive it."

"Yeah, and how's a lard press going to help us survive?"

"It is a very fine lard press and one can not buy one just anywhere! I was lucky to find one. You will appreciate it when you need it some day."

"Well, you do come to appreciate all sorts of odd items on the homestead. If not for their original purpose then for the parts. But about your outhouse. You could use your tent later when you move into the cabin. But meantime you could use that canvas tarp you have and make it into a tent. Just dig a hole where you want your outhouse to be, build a box with a toilet seat or board lid to cover an appropriate sized hole in the top, then set the box over the hole you dug. Just make sure you don't dig your hole too wide. Then arrange your tarp tent over it all. I'm sure you can come up with something to hang over the ends for doors. That would do until you can build a more permanent structure."

"I knew it, I knew you would be coming to the building part. Every project has me building something. This old arm only has so many hammer blows in it, and this old body's not getting any younger you know."

"No, it's just getting healthier. You're not that old, JJ. You probably have a few more projects left in you. After all Helen and Scott Nearing were 50 and 70 years old when they moved farther north to begin their *second* homestead. And a toilet box seat isn't that big of a project."

"Sue is right, JJ. We have plenty of boards left from building our cabin. We should be able to come up with a fine box. Besides, you are not the only one sawing and nailing. And it *would* be nice to have a structure over our outhouse, even if it is thin. Yes, I think that is what we will do. We can take time off of the cabin to build a box and put together our tarp/tent/outhouse. And we will build a wooden building later. We can do that this afternoon, JJ. And you will no longer have to be a bear lookout in the woods for me."

"Well, I suppose we can build an outhouse box seat all right. But I'm not digging the hole until you decide exactly where you want it. I don't want to be putting holes all over the yard like when you couldn't decide where you wanted that apple tree seedling."

"It shouldn't be too hard to find a good spot for your outhouse. This area is good deep well drained sandy loam soil so you don't have to worry about drainage. And there is not a creek or lake or river to stay away from. Though you will want to stay away from any water source, such as your well. Somewhat centrally located for summer and winter activities is nice. And it's nice if your door looks out onto a pleasant, private area. In our case that is the woods.

Although a well managed outhouse doesn't smell bad generally there are times of the year when it may have a little stronger aroma than usual. You'll want to consider that. A vent pipe out of your box and the generous use of wood ashes and sawdust or similar materials keeps those possibly offensive odors to a minimum. You won't have to worry about ventilation in your tarp/tent outhouse but you may want to think about it for your next structure. We get enough with a loose fitting back window and door.

You'll also want to make sure you have easy access for emptying. Once a year, usually in the fall, I take shovel and wheelbarrow to our outhouse and dig out the well composted biffypost from the side that has been sitting for a year. This is spread under the fruit trees and around the berry bushes and plants. The seat box is moved from the current side to sit over the now empty hole. The current in use, now full, side is covered with ashes, sawdust, and dirt, then boards laid over for flooring. That side will sit and compost for a year to be ready to be dug out and spread the following year. It works great.

Previously we had one hole with a scrap of plywood dividing the two sections. And the box just moved from one side to the other. But when our outside work schedules changed, and we were working at home more, it made quite a difference in terms of, well, quantity of biffypost. The divided hole design wasn't big enough. So we dug another hole near to the outhouse and using come-along and rope moved the whole building over the new hole. We haven't yet decided if we will continue to move the building back and forth each year, build a second outhouse, or just build a new larger one. You'll want to think ahead to what your future needs might be when designing your facility.

We do have an indoor toilet as well. We put in a septic tank when we built our house. But once you're used to the airiness and view of a nice outhouse an indoor toilet is rather, well, stuffy and cold. And the practicality of compost versus a dead end septic tank is a strong incentive for the use of the outhouse. Your body soon adapts to fast visits when it's cold out.

An indoor composting toilet is also an option. There are commercial models as well as a limited amount of information on building your own. This is something we've thought about but just haven't done yet.

If you don't have an indoor toilet a bucket with a lid works fine for night time or day time not-wanting-to-run-outside-in-the-cold-or-rain whizzing. For kids or adults. Just empty it often onto the compost pile or an unused garden bed. A good rinse every time keeps it odor free. It helps if it's in a cooler part of the house too. By the way, a piece of dense foam cut to fit on your outhouse seat (or some such insulating affair) is a standard, and highly recommended, accessory in most north country outhouses.

The aesthetics of the outhouse should be mentioned also. I've never seen an indoor bathroom that was anywhere near as creative and memorable as many outhouses I've visited. Sometimes it's the view; a mountain, a distant lake, a wilderness woods, a field where any and all sorts of wildlife may come wandering or flying by. Sometimes it's the location such as on the side of a steep mountain or the top of a tall wooden structure (just open a door at the bottom to clean out at ground level instead of digging out of a hole). Or it's the decor. From rustic to baroque, plain to painted to wall papered. And the wall paper is seldom boring. Most outhouses are interesting, which is a word you seldom think of in describing the typical indoor job. I have no doubt you two will come up with a unique outhouse to fit your homestead selves when you get around to building one."

"I believe that a tent outhouse is quite unique enough for right now. But I shall appreciate it all the same you can be sure."

"I can just picture our outhouse, CindyLou. Shag carpeting for the feet and a fireplace for the body and a well stocked library and an attached sauna and . . ."

"We are a homestead not a ritzy big city motel, JJ! We do not even have a door or a bed or a closet or a couch yet in our cabin. Or a roof for that matter! And if we have to wait for you to research and think and design and draw plans for an outhouse like that I will still be having to go into the woods this winter when the snow is over my head!"

"OK, OK, unruffle, unruffle! We'll just get together our tent outhouse for now. No reason it can't have some necessary amenities though. And it doesn't snow *that* much up here. Does it? Maybe we should have built a taller cabin."

"Don't worry, we'll be glad to come over and dig down to get you out. We have good shovels."

"You are kidding are you not?"

"Honestly, CindyLou, the snow has never been a real problem for us. Believe me, you'll handle it when it gets here. If you can deal with hurricanes you can deal with snow."

"I suppose you are right. I will have to take your word for it. But it is a beautiful day today and it is not snowing. So let us go, JJ. We have a biffypost hole to dig! And I know just the spot to put it. Thank you for the strawberries, Sue. Hurry up, JJ, the sun will not wait for you and we have a homestead to build! *Ohhhh, we'll make us an outhouse, a pleasure to seeee* . . . and it will be truly a pleasure to have when it is done, JJ. I wonder how bear-proof a tent is?"

Breakfast Flowers and Luncheon Weeds
Eat What You Grow and Grow What You Eat

~ ~ ~

"Hi there JJ, good morning CindyLou. I'm over here behind the beans. Did you come over to pick breakfast?"

"If I was a goat I could graze my way through, I'd have better luck finding my way. But I assume you're kidding about the breakfast. A good breakfast comes out of a box not a jungle. That is, it should. If CindyLou hadn't forgotten to buy any. A man's got to have food to get him through the day you know. Especially with all the work I've been doing. Why just yesterday . . ."

"Good morning, Sue. Watch your step, JJ. You stepped on that vine and just about trampled that squash. Look out for those beans! Sue is not going to let us in her garden if you are going to step on everything. It is a good thing our bed is on the outside edge. And I did not *forget* to buy cereal. We are homesteaders now, we do not eat out of a box. We have a garden. We will eat out of the garden. There is a lot there to eat. I am sure there is. We just have to figure out what it is and what to do with it. It did look a lot neater when I planted it. And I do not believe you are on the verge of starvation yet."

"Now, CindyLou, I'm not knocking your garden bed. There is sure enough a lot growing here. If you were a cow or a chicken or a sheep or something. But a working body needs real food. Darn, sorry about that. You ever think of putting paths in your garden, Sue?"

"Well, we do have them but some of the plants tend to spread out this time of year. Come on over here by the onions, they don't sprawl. I'd rather grow food than paths I guess. It's such a short time that we have a lush garden I don't want to discourage anything. Some years I do a better job of staking things like the tomatoes. Then others I get in a hurry and just let them go. They don't seem to care all that much. Unless it's a particularly wet season then I do try to get everyone up off the ground. The mulch helps.

Your garden bed is growing great. But I think you'll find a little organization doesn't hurt, especially when it comes time to harvest. But don't worry, I'll help you sort it out. See, here's a carrot, and this cucumber comes from over there I think, and look at the tomatoes on this vine, they'll be ripe soon. You might want to scooch them out from beneath that squash vine though. This is a garlic and here is a potato plant. I'm not sure what this is, doesn't look familiar. You might want to consider doing a little thinning, CindyLou. Things won't grow to good size if they're too crowded."

"I know, you told me that last month. I do hate to take anything out. It is our first garden and I want to eat out of it. And I am worried about pulling out something edible even when I think it is a weed. But I do not remember planting so much. Do you know what this is? Or this?"

"Well that one you can pull out, though you can eat it too, it's a good green. It's called Lamb's Quarters. But it's crowding that lettuce. Maybe cut it off next time you want a salad. That one there is edible too but you definitely want to get it before it spreads. It's good to eat, called Wild Sorrel, but it is a real pest in my garden. Take my word for it you don't want it to get established anywhere. I don't recognize this, or that one there, looks like they're some kind of flower. I'd weed them out. They may be pretty in the fields but they're a weed in the garden."

"Flower? There are some flowers growing there? Hey, CindyLou, don't pull that out. I don't believe it, they actually grew! What do you know."

"What I do not know is what you are talking about, JJ. I did not plant any flowers in our bed. And this one, whatever it is, is shading the onion."

"Now just leave him be, CindyLou, he isn't going to hurt anything. I guess those are *my* flowers. Maybe this gardening isn't all bad. Though it's not filling my stomach you know."

"I am not worried about your stomach, JJ, it will do it good to be empty for a minute. What are you talking about the flowers? Why are there flowers in our garden bed? We can not eat flowers and I did not plant any."

"I did, CindyLou, I did. There were some pretty packets of flowers on sale cheap at a store I was in and I just bought a whole bunch and spread them all over your bed here. Thought they'd be real pretty. But they didn't do anything. Except for those couple there, if they really are flowers. They don't look much like the pictures."

"*Our* garden bed, JJ, our bed. And I am sorry your seeds did not grow, even if I do not think we need to plant flowers. I wonder why they did not? Maybe there was just not enough room."

"It is rather crowded in there, CindyLou. But there are probably several reasons why your seed didn't grow,

JJ. There's a good chance that what you bought was old seed. And while most seed stays viable a good number of years if care is taken, those seed packets probably weren't in that category. And many flowers need to be started indoors early and transplanted out in the garden. With our short growing season they just don't have enough time to mature and flower if you plant them direct. And just throwing them on top of the garden bed doesn't give them the best chance to survive and thrive even if they are short season varieties. Most seeds need a little more care than that. When you get your garden plot you'll have enough room to plan and plant a little more carefully so everything has a chance to grow."

"Well, it sure is a large plot CindyLou has me working up. Took me practically all day to till that buckwheat in yesterday! And what with the bees chasing me and the sun beating down on my head I nearly expired from the exertion!"

"It did not take you all day, JJ. And it was a cloudy day yesterday. And the bees did not bother you that much, I was watching. You were killing their flowers after all. I am surprised they were not more angry. And it is not too large of a plot when you think we have to grow all of our food. And now you want to plant flowers too. You are the one who is worried about breakfast. Now do you want that flower or that onion? You can not have both."

"I wasn't just thinking about having either for breakfast, CindyLou. Now don't you go pulling out that flower. You can get onions in the store."

"That is not the point, JJ! We have a garden plot now and we are going to eat out of it. Do you want breakfast or do you not?"

"Well, CindyLou, your crops *are* maturing but I don't think you can expect to get every meal here. You certainly can eat entirely from what you can grow and gather but it takes some planning. And a larger garden. But for now I think you'll want to supplement your garden produce. We'll be glad to share some with you though. Come on over here, there are some Early Chatham tomatoes ripe. Have one for an appetizer, JJ. Here you go, CindyLou. Now, what were you thinking of harvesting?"

"I suppose you are right. And thank you. I do not know what I want to harvest. The things we are used to eating do not grow in the garden. I do not even know what to do with what I planted. But I do want us to eat better. I just do not know what to eat. But this tomato is *very* good. I did not know tomatoes tasted like that! We must have some of these in our garden, JJ. Maybe tomatoes are what we could have for breakfast."

"Well, now, this is a right delicious tomato I have to admit. But I'm not sure that's what I want for breakfast. Besides, we could get awfully hungry waiting for these things to ripen. Ours aren't even red yet. And if we eat them for breakfast how are you going to make salsa? I have my heart set on some nice good salsa this winter. You do have hot peppers growing out here somewhere, don't you, Sue?"

"JJ!"

"That's OK, CindyLou. Sorry, JJ, no hot peppers. Peppers are difficult to grow here, though I usually put in a short season variety or two. They're not doing too bad this year, but they're sweet peppers not hot pepper. We should have some extra tomatoes though if the frosts don't come too early. But a little later you'll be able to buy a lot of produce at the farmer's markets and from people who grow for sale. It's a great way to get your supply of food while you're waiting for your own garden. Or to get things you can't grow, or if you had a crop failure. It's not the same as fresh from your garden but it's the next best thing."

"Well, how do you decide what to plant? And what do you eat? I know we have shared meals with you but it is hard to think of how it gets from the garden to your kettle. There is so much to learn."

"And breakfast is going to turn into lunch which is going to turn into dinner if we don't get something decided here soon. I knew I shouldn't have let you do the shopping this week, CindyLou."

"There are some raspberries over by the fence that are ripe. We can talk while you get your breakfast on the hoof. So to speak."

"I'd be a lot happier having it off the hoof in my frying pan. OK, OK, unruffle there, CindyLou. Lead me on to the raspberry herd. Guess I've done weirder things, and the old stomach's not exactly in a position to quibble."

"Now, let's take your questions one at a time. What to plant is a personal decision. But it's simple. Just plant what you want to eat *And* what will grow in your garden. To begin with just put in whatever you think you might want. You'll soon find out what you eat and what you don't. The next year don't bother to plant what you didn't use the previous year. If you have a small plot then you'll want to take best advantage of every inch of space. You'll probably not put in much corn, or sprawling squash, or beds of dry beans. Unless they are very important to you and worth taking up limited space. If you don't like something don't plant it. If you eat beets then you'll want them in your garden. If you don't eat them then there's no reason to plant them. Unless there is someone you want to give them to of course.

Harvest time is a consideration too. Both Steve and I like fresh peas. But when they are ready to harvest we're often on the road and very busy with other activities. So after many years of missing the harvest I finally decided not to plant regular green peas. I just plant a small patch of snow peas since they produce over a long period and can be eaten from the time they are small as pod peas until they are full size as green peas. And what I don't harvest green I leave to be harvested for dry soup peas. It's a trial and error process to adapt your garden to your own likes and needs.

It's surprising what you don't remember one season to the next so be sure to keep notes. Keep a notebook or clipboard with your garden tools and use it often. This will make it easier the following years when planning your garden. Don't be afraid to change your plan to fit your changing lifestyle. And the changing weather. If you get too hung up on having to grow a particular crop or knowing you have to plant on a certain day you're sure to be disappointed. But if you stay flexible you can be happy with what you have all of the years.

Don't limit yourself to just what you know either. Or what you've eaten in the past. Be adventurous! You never know what you might find that you like. Go ahead and try new things. Just remember to actually harvest and eat what it was that you planted. You might find some new crop to be quite good, but not worth the trouble. Or maybe it is a great grower but you just don't like the taste. Or maybe you will find the perfect new vegetable for your table.

Let your breakfast, lunch, or dinner be different. We often get carried away with the lists of foods we *have* to eat at this meal or that one. Common sense is a better companion than an arbitrary list. The animals don't eat four or five course meals at preset times every day. And they do a better job than we do in the health area.

Some people are more adventurous in the kitchen than others. Just depends on how much you enjoy cooking and experimenting, and how much time you want to spend doing it. You can have good food with a lot of hours or with few. Neither Steve or I have much interest in that area. No one accuses us of being great, or even partly great, chefs! So our meals tend to be simple. They change with the seasons, as they change with the years. Our work schedules also dictate what we eat, and when. It's so variable it's hard to make generalizations. But I understand your question so I'll give it try.

Since breakfast is on your mind I'll start there. Cooked oatmeal used to be our usual breakfast meal all year round. We like it lightly cooked, not overly mushy, with chopped up apples (dried or fresh), or berries, and sunflower seeds or raisins if we've chosen to buy them. The crowning touch is maple syrup. Steve's the breakfast cook. I do lunch, and we more or less take turns on dinner. Though when Steve's working out I get dinner together on those days too. And we stay flexible always if one or the other is tied up in a project. Now oatmeal doesn't grow in our garden but it is an inexpensive food and one we appreciate being able to buy. We get organically grown oats in bulk from the food coop.

One day Steve came up with a new breakfast which has been our mainstay since. Three scoops of raw oats in a bowl (about 3/4 cup), add fruit sauce, a little water if you want it thinner, maybe some sunflower seeds. Top it off with maple syrup, stir, and there you have it. Homestead instant breakfast. The basic menu stays the same pretty much year round but the fruit sauce changes with the seasons, starting in early spring with the first rhubarb. That first fresh rhubarb sauce is just the thing to bring in the growing season. We also use a lot more maple syrup that time of year as rhubarb sauce is pretty tart!

When the strawberries start coming on I add those to the rhubarb sauce, until at the height of the strawberry season we're eating straight strawberry sauce. Then back to rhubarb/strawberry mix until the strawberries finally produce themselves out for the year.

Then the raspberries ripen, and the blueberries. How often they end up in the breakfast sauce depends on how much time I take to go pick them. I used to have tame raspberries and blueberries growing in the garden, but they were often neglected because of the immediate demands for attention from the annual crops. One year I realized that it didn't make much sense to work so hard trying to grow those crops in the garden when we live in the middle of an area rich in wild berries. So I took them out. Though not as convenient, it is a lot easier to let Nature grow those crops for me.

I also dry rhubarb for winter use, and strawberries when the crop is good. Sometimes I'll can up some blueberry or strawberry sauce, too. Later in the season we have apples to add to the rhubarb sauce. And as the rhubarb gives out we end up with plain applesauce. Now, we're in apple country here so apples are an important crop to us. I can some and dry a lot. Dried apples are great for simmering up into late winter/spring sauce and for snacking anytime.

In the late fall we store several bushes of apples in the root cellar, buying from the local orchard if our crop is not good. I make fresh sauce from the apples well into the winter, adding dried or canned rhubarb, strawberries, or blueberries. When the fresh apples are gone we use the dried apples. Then later when it is warming up and the wood stove isn't going much, and the solar oven isn't out for use yet, we use canned sauce.

We still cook up oatmeal when the mood strikes us. Or something else if we want, such as pancakes, or potatoes and eggs. But the oatmeal/fruit sauce breakfast is fast and delicious. So that's our usual fare.

Lunch and dinner depend on what is ripe in the garden, if that time of year, or what's in the root cellar and pantry. Dried beans and peas are a big part of our diet. Whenever we have the wood stove going there is usually a pot of beans or peas cooking. And in the warm months, when it's sunny, I use the solar oven to cook them. Since they take several hours to cook I don't cook that kind of meal with the propane stove. Add potatoes, maybe carrots, dried green corn when we have it, sautéed garlic and onions, a few herbs such as sage and celery, some dried or fresh cutup greens, a splash of homemade chokecherry wine, a touch of pepper and salt. You can't get much better than that!

Dried beans cooked with dried green corn, celery,

herbs, sautéed garlic and onions, and tomato sauce is another great winter meal. Maybe add some cooked macaroni for a change.

In the summer beans and rice are a common dish. It's so easy to stick them in the solar oven to cook in the afternoon for a good hearty dinner that evening. It's perfect when you're busy. Since the beans take longer to cook than the rice I cook the two separately. Sometimes I'll add other vegetables towards the end of the cooking time, though beans and rice plain are good too.

When the broccoli comes on we have broccoli soup with potatoes and the usual garlic and onions (we're rather big on garlic and onions as you may have noticed). Broccoli is also good stir fried/steamed, along with rice or potatoes. Earlier in the season we have asparagus the same way. We eat asparagus until its season is over and we're getting rather tired of it. But by the time next spring comes we're looking and waiting anxiously for those fresh new spears to appear!

When the tomatoes are ripe on the vine, which is a very short time for us, we relish fresh tomatoes. Cut up and added to any other fresh vegetable you can eat them cold with macaroni for a salad, or cooked with rice or potatoes. Or sliced onto an onion and pickle sandwich, with cucumber if you like. Mmmm. Maybe that's what we'll have for lunch today.

When the corn is ripe we have green corn on the cob. When the peas are green we have peas in our meals. Snap beans when they're in season. Potatoes from mid summer through spring. Carrots summer into winter.

Winter squash makes a good light meal, with biscuits or toast. And the leftovers are great as squash soup, with onions and garlic and some dried greens. Cooked squash is also a common ingredient in my cookies from fall into winter.

Potatoes and onions are a good quick meal almost anytime, with fresh or dried greens, and maybe eggs.

In the summer I don't do as much baking but bread and biscuits and homestead cookies (which aren't much like what you usually think of as cookies) are regular cold weather fare, when the wood stove's going and it feels good to be baking.

The one canned item we buy is tuna fish, both for us and the cats in the winter. So we have an occasional not-so-homestead meal of macaroni and tuna salad, with onions, greens, pickles. The salad dressing is a mix of vinegar (usually homemade wine vinegar), oil, herbs, and pickle juice (gives it some sweetness). This dressing is good on potato or macaroni salad, or for a regular green salad.

And there are times when we just aren't very hungry and a meal might consist of popcorn, with nutritional yeast as seasoning, or pickles and rolls, or homestead cookies and peanut butter. Or carrots, maybe with peanut butter. From the time the first carrots are big enough to eat in the summer they are on the table almost every day during the growing season.

In the summer we may have a salad of whatever is green and growing in the garden. Though I've stopped growing lettuce and most greens because it seems I'm never there to pick them when they are ready to be picked. But I know for many people the salad garden is top priority. If I want a green salad I use garden sorrel (a perennial) and swiss chard leaves, plus maybe lambs quarters or milkweed. Swiss chard is easy to grow and it is there all summer and fall for either salad or cooked greens, without overgrowing or getting bitter.

And speaking of cooked greens, that's a good meal too. Sauté garlic and onions, add the greens, then a bit of water and let them cook until limp but not mushy. Cutup apples are a good addition, as are sunflower seeds.

In a good healthy garden there is no lack of food to eat. The question is more what to choose than what is available. Our garden is probably simpler than most as far as variety of crops grown but, as you can see, there is no lack of good meals to be had. Mostly it's a matter of habit, of getting used to going to the garden or to the pantry or root cellar for your food instead of to the grocery store. Once you decide you are going to eat close to the earth the biggest challenge is over. From there it's simply a matter of doing it.

Even if you don't have a garden yet, you can eat well, inexpensively, and heartily by buying in season and in bulk from those farmers and gardeners around you. Add a few staples from the food coop and you're all set. Instead of trying to duplicate the same foods you are used to buying in packages, start fresh with what you have and what you can get to come up with new meal habits. Experiment! It'll soon become a habit and one day you'll go into a grocery store and not be able to find anything to eat! That's a major obstacle for us when we're on the road.

I haven't found many recipe books which deal with eating from your garden without adding packaged or store bought food. But one which I can highly recommend is *Mrs. Restino's Country Kitchen - The Complete Wood Stove Cookbook* by Susan Restino, the 1976 edition. There is a newer 1996 edition which is nice but not quite the homestead bible that the earlier one was. It is still a good buy though if you can't find a copy of the 1976 book. My copy is quite worn from use. The classic *Joy of Cooking* can be a good tool also. There's a lot in there you'll never use but they also include good, plain cooking advice using good, plain food.

But you really don't need recipes. Just go into your garden, pick what's ripe and have it for a meal, cooked or raw. Or select something from the root cellar, or pantry shelves. Give yourself a chance to really find out what you're eating, without drowning the food in anything else. You'll soon be thinking of what might taste good with what, and trying a little of this, or a little of that with that dish. Or experimenting with one combination, then another. Before you know it you'll have more ideas than meals to cook."

"Well, I can't say as you've convinced me that tomatoes and raspberries make the same meal as steak and eggs. But I guess a body can get used to most anything. Or so I'm finding out. Though I'm not sure a body can sustain this kind of life on that kind of food. I just never heard of great muscles being fueled by pickles. And these here muscles need all the help they can get. CindyLou keeps coming up with more uses for them than you have uses for these crops here. They don't last forever you know, and at this rate . . ."

"Well, you have given us some very good ideas, Sue. I am still not sure what to fix for dinner. And I do not think I am ready to turn that entirely over to JJ. But I think I can see where to start. I am just not sure where it will end."

"You can always buy a block of cheese, CindyLou. That seems to be an easy pick me up for most any meal. It's good at hiding some of the more . . . different experimental meal results too. And it might be a good bridge between the old and new ways of eating. I've found cheese lasts longer if you take the plastic wrap off and wrap it in a clean cloth instead. And keep it in a cool spot."

"What about pickles? I have never made one. In fact, I never before thought of them being made. Where do you have them growing?"

"Pickles aren't hard. I'll be glad to share my recipe with you. My cucumber crop is over here. That's what you usually make pickles from. But when I don't have a good cucumber year I often use immature squash. They make good pickles too. I used to grow zucchini just for that purpose since that seemed to be more reliable than my cucumbers. But because of my seed saving I don't grow them anymore, don't want zucchini crossing with my winter squash. It looks like there are enough cucs ready right now to make a batch of pickles. Come on over this evening if you want and I'll show you how I get them going. We'll get them started tonight then can them tomorrow night. Because I use the wood cook stove for canning I like to do them first thing in the morning or at night. We don't mind the heat from the stove as much then."

"Well, I'm all for the cheese. At least I'm on speaking terms with that food. How about I run into town for some right now, CindyLou? Maybe pick up a few things while I'm at it."

"No, JJ. We decided we would live off of the land when we were out of the food that we had bought before we moved here. And that is what we will do. You did say you wanted to do that too, JJ. But maybe we could buy some items from your food coop, Sue. And maybe we can think about some cheese too. But no prepackaged food, JJ. We can do without that. I know we can. At least I think we can."

"I know I said I would, CindyLou, but that was before we ran out of spaghettios and cheerios and potato chips. And I don't care what we decided, a person can't live off the land for more than a few days in this day and age. It just can't be done. All that food self-sufficiency stuff is a lot of crock and PR. A person needs *real* food to survive."

"Well, JJ, I hate to disagree with you on such a beautiful day. But how about if I lend you a couple of books. I think you might find them of interest along that particular subject. Two are by Helen and Scott Nearing, *Living the Good Life* and *Continuing the Good Life*. In fact, didn't I already send those over to you, CindyLou?"

"Yes you did. We have just been so busy with all of our projects that we have not read them. I did look through them but JJ has not. From what I saw I think you are right. I will make sure JJ gets them. What is the other book? Do we have that one?"

"You might, it's called the *No-work Garden Book* by Ruth Stout and Richard Clemence. I think you will both like that one. Another book by Ruth Stout is called *How to Have a Green Thumb Without an Aching Back.*"

"Now that sounds like my kind of book. I like her already. I keep telling you, CindyLou, this life just isn't as easy as it should be. I should have known someone would write about how to make it simple. Can't wait to read them."

"Well, JJ, I think you might be a bit surprised by what they say. But I'll leave that for your discovery. As far as being able to subsist off the land in this day and age, you can. Many, many people have done it and so, as a matter of fact, have we. How about I tell you about that this evening while we do the pickles."

"I will be happy to hear about it. And will be happy to learn about pickles. Thank you. And now that breakfast is over, JJ, we must get back to work. You did say you did not want those muscles of yours to get out of shape. So let us go exercise them. We have a cabin to finish. See you later, Sue."

"OK, thanks for stopping by."

"What do you mean breakfast is over? That was fine and good for a start but you can't actually think that a few tomatoes and raspberries are going to sustain this body all morning do you? And I did *not* say I needed to exercise my muscles. I am sure I didn't say that. Hey, wait up CindyLou. Those things in my stomach just shouldn't be jostled too much I just know it. Why, if you don't take care of your stomach, do you know what can happen? I was reading just the other day . . ."

~ ~ ~ *eleven* ~ ~ ~

Pickles and Eating Out . . . of the Garden
An Adventure in Self-sufficiency

~ ~ ~

"Hello, Sue, I am ready to help you to create pickles. I even searched and dug deep to come up with what I consider to be the right pickle outfit."

"Hi, CindyLou. I was just getting ready to pick the cucumbers. And that is quite a dandy, um, pickle outfit. Quite the bright green color. Is that what makes it a pickle outfit?"

"Yes, I thought the color a great pickle color. Even though we are beyond the shopping centers I feel I should not let my fashion sense be left behind. There is no reason one can not have ruffles in the country. One does not have to stoop to flannel and T-shirt in order to be a homesteader. Unless that is what one wants to wear of course. I did not mean . . . that is . . . your T-shirts are often quite nice. And I am sure those pants are, well, they must be . . . comfortable anyway. You do not go shopping very often do you?"

"Not unless I'm roped, tied, and blindfolded. Don't worry, CindyLou. I'm happy with my clothes and you're happy with yours. That's all that counts. We certainly don't have to dress alike to be friends. Whatever works for you is what's best for you. And I find your outfits quite . . . interesting. Now, if you want to grab that bucket there you can pick the cucumbers on that side of the row. Just the ones that are about three inches or longer. Be careful pulling up the leaves and vines to get them, they're prickly."

"Why not pick these little ones, they are very cute. Ouch. Picky little critters. I can not believe they are going to be edible. Are you sure these are what you make pickles from?"

"Sure enough, CindyLou. Just don't grab them very hard. The little spines mostly scrub off. And the cucumbers soften up in the pickling process. You can pickle the little ones too if you have enough cucumbers and that's what you want. At the end of the season I pick them all down to the tiniest. But as long as frost isn't threatening I leave the small ones to grow large. I want slices instead of whole pickles. If you process them soon after you pick them you won't have to worry about mushy pickles. They'll be crisp and good. Did JJ come over with you?"

"Yes, he went to find Steve. He had some ideas for the latches for our cabin windows that he wanted to ask him about. It really is coming along. I can not believe I will be living in my very own place soon built with my very own hands. Or rather our very own hands. I think it is quite a grand place. Not that I do not like our tent. But it will be very nice to be moved in. Even if it will not be quite done. Though it should not take long to finish it once we are moved in. Before the snow comes down I am sure."

"Mm. Well, don't worry if it takes a bit longer than

that. You have a lot going, and it will be quite enough just to get moved in. There, I think we have them all. Let's head in.

"There really is not that much to this. Just scrub the cucumbers well, slice them up and set them to soak in a salt water solution until morning. Hi there, JJ. Get your latches figured out?"

"Yep, think we did. They're going to work just fine. Steve gave me some ideas on how make them. He's just finishing up what he's working on then he'll be in. Thought I'd come over to see what you two were up to. And get a first hand hear of what kind of projects you might be cooking up for me to get involved in. Not that I don't have enough already. And maybe get a drink of water while I'm at it. Been quite a thirsty day today."

"Help yourself, JJ. CindyLou tells me you're really coming along on the cabin. We'll have to walk over and see how it looks. And I'll try to keep the project ideas to a minimum tonight."

"Project ideas keep a person young and alive, JJ. They will not hurt you any. And I have not added *that* many lately. Besides, Sue was going to tell us about being food self-sufficient. That is not a project. At least one does not have to build it. I assume."

"It's more a philosophy I think, CindyLou, than a project."

"And it's a philosophy that's just in the books, not in real life. One can't live on hay and turnips you know. You can't be food self-sufficient without starving."

"Actually, JJ, people have been feeding themselves off the land for eons before us. And the animals do it all the time. You just need to fit your wants and needs to your abilities and resources. Not to say there aren't places where you'd probably go hungry. But that's not here."

"Hay and turnips. CindyLou's going to be feeding me hay and turnips, I can just tell."

"JJ, I do not even know how to grow a turnip. And

I have never eaten a turnip. I did not say we are going to be eating turnips. I just said that we *could.* And did you not listen to a word Sue said this morning? She was not talking about eating hay and turnips. And besides, many creatures are very happy to eat hay. And many homesteaders eat turnips. Along with other things they grow. We are homesteaders and we can be self-sufficient in our food. I read it in those books you brought home from the library. They almost always write about being food self-sufficient. They say you have to be food self-sufficient so we will be. We are going to be good homesteaders, JJ."

"Yeah, and while they're writing all that they're sitting in a restaurant chowing down on a big old plate of lasagna and steak with a gigantic chunk of pie and ice cream coming for desert. Oh, does that sound gooooooood. Admit it CindyLou, you can't live off the land. Even Steve and Sue buy some of their food and they have a gigantic garden. Now, let's head down to the corner for a bite. Those pickles look pretty near done. What time does Tylene's close anyway, Sue?"

"JJ!"

"Think they close about 9:00 this time of year. And while we make good use of that restaurant ourselves, and enjoy doing so, you *can* feed yourself from your garden and the land around you if you want. With or without hunting game. And the pickles aren't quite there yet, JJ. Sorry.

And yes we do buy some of our food, things we can't grow, things we choose to have. But that doesn't mean you can't live quite well on what you can grow and gather. The biggest obstacle is simply thinking you can't. All you have to do is decide you want to. Our habits and perceived needs get in the way more than anything.

I do agree with you though, JJ. We tend to get a little glib when talking about being self-sufficient. It's usually in terms of an overall point of view as opposed to particular fact. You don't have to do absolutely *everything* yourself to live a self-sufficient lifestyle. Few people would want to. I know I wouldn't. It's more the idea of being responsible for yourself and your part in the world. As an integral part of that world, not separate from it.

But the challenge and fun of eating only what you grow and gather is a good one. And it can be fun. You don't need to subsist on a strange or deprived diet either. I think you two would come up with something other than turnips to eat! And as I said, feeding yourself isn't new or unique. People have been doing it for ages. The idea that food comes out of a box or can is a modern phenomena. And I think it would be a strange and funny one to many people in the world today.

Aside from the fun and challenge of subsisting through your own hands it's great to know that you *can*, even if you choose not to. And it gives you a great appreciation for those things which you can't grow or produce yourself.

A number of years ago we decided to try that very thing. To eat for a year, when we were home, only what we came up with from our homestead and surrounding areas. Since we keep no animals and are not hunters that meant what we could grow or gather from the wild. We started thinking of it early, in the winter, when planning the garden.

Since this was an experiment to use in our lives, not just an abstract something to write about, we made compromises and practical decisions. We knew we could survive with whatever we came up with, but we wanted ideas we could incorporate into our lives every day. I planted the garden with our goal in mind. We would start in the fall since summer is quite a chaotic time for us and we are often on the road a lot. And when home we were already eating mainly out of the garden.

That summer proved to be a rough gardening one, with cold and wet and frosts. We had a lot of blight in the potatoes but there were enough for the root cellar.

The tomatoes had more blight and rot than usual, but the jars of sauce multiplied on the pantry shelves. Thankfully, we had a decent dry bean harvest in spite of the weather, and the cole crops and greens thrived. The dry peas likewise. And the carrots and onions made a decent showing. We found we didn't eat turnips even if they grew well. And lentils are not very practical in our climate and in a garden setting.

Grains were the biggest problem. I grew quite a bit of wheat and barley and millet, and the harvest, while not the best, was not bad. However, the small bucket of grain threshed out was hardly a week's worth of food, let alone a year's. You need quite a patch of small grains to supply even a small family's needs, and you need to be able to harvest, thresh (the most difficult part) and winnow it. A tall order for a small garden, and not very practical.

Corn makes more sense. If you can find a short season variety. And if you can keep the raccoons out. The varieties I had that year were OK. But the coons weren't. Actually, they had a great time. But they left very little corn for us. So we learned. If we were truly feeding ourselves from our garden we would be eating very little small grains. And we would have to keep the coons out of the corn. Which we have subsequently been able to do, as well as come up with a short season corn, so I can provide much of our grain needs in the form of corn now. But that year we bought wheat and corn meal. Keeping in mind in our "eating from the land adventure" that we would normally not have much wheat to eat.

Another lack we ran into that year was one we normally didn't have - sweetening. We tap our maple trees and use our own syrup for all our sweetening needs. I'd learned to use maple syrup in jam and pickles, cookies and tea. But that year our firewood supply was low and we didn't cook down as much syrup. We were going to come up short. So we bought honey to supplement what syrup we had.

And then there was oil. We came to realize just why it was such a prized and valuable commodity in years past. Whether it was bear grease in the fur trade era or olive oil across the globe. We could grow sunflower seeds, and we did grind some up, floated off the hulls, used the wet ground seeds in a biscuit. It was OK, but not very practical for any volume; either growing, harvesting or processing. We did have hazelnut bushes around, a theoretical source of oil. But the chances of having a large enough harvest that there would be any left for us after the small furry critters were through was quite small. About one year in seventeen so far. And we're talking about a few quarts, not shelled, at that. Beechnuts? Have you ever seen the size of a beechnut? Nuff said. Conclusion: no oil.

Well, we proceeded with the experiment. And we lasted about two and a half months. It was a great learning experience and our eating habits did change as a result. We buy even less of our food now than we did before. More important we found out that yes, we could, if we needed to, if we wanted to, feed ourselves entirely from what we could grow and gather. We would have to take more time doing so than we normally do, and it would have to be a high priority, but it can be done and you can eat quite well.

As we added outside purchased food back into our meals we did so because we consciously wanted to, not just because of habit. We found out what was most important to us, the things that come first out of the food budget. The first two things we added were oil and oatmeal.

We found one can live quite well without oil, but it sure is nice to have. You can pop popcorn by heating a heavy pan well, pour in the popcorn then a generous splash of hot water. Have the lid ready and take care the steam doesn't come back to burn you. You can cook vegetables (steam instead of sauté or fry). You can make breadstuff and cookies (make biscuits and flatbreads, and take the cookies off the pan while still warm). It's not the same, no, but it won't kill you. Doesn't even hurt that much.

Lunches and suppers weren't all that different. The biscuits and cookies weren't the same but not that far from our usual fare. But breakfast was a challenge, because of a long, well ingrained habit of oatmeal. We tried cooked wheat for awhile. You had to remember to get it cooking the night before but it was OK for a breakfast cereal. However, neither of our bodies could handle that much wheat. And if we were growing our own we wouldn't have that much anyway. Corn meal mush was an option that didn't catch on. But we could get used to it if we had to. We ended up with warmed up leftovers or potatoes and onions. Both were fine. But we missed the oatmeal.

We limited our wheat use during our experiment and learned to use more cornmeal. I found that to use cornmeal in baking it helps to soak or even cook it a while first in water. That's not so important if you have flour corn to grind but flint or dent corn makes a courser meal. And cookies, biscuits and flatbreads don't hold together as well without wheat flour. You have to use more moisture and make the flatbreads a little thicker. Again, flour corn makes these easier.

Assuming a shortage of flour if we were providing all of our own I learned to use other ingredients. Most dried vegetables can be ground up for use as flour. The easiest being dried greens such as swiss chard, which is quite prolific in my garden. It has no sticking power but it has bulk. Another ingredient I learned to use in biscuits, flat breads, and cookies is mashed potatoes or squash. Other vegetables would do as well I would think. Just add whatever flour you have to the cooked, mashed vegetable to make your dough and proceed. Applesauce, or other fruit sauce, can be used in the same way to make a cookie. Add flour and sweetening, drop on a cookie sheet, flatten with a wet spoon and bake.

There were several purchased foodstuffs that we found we didn't miss that much, such as raisins, which used to take up a good chunk of our food budget. And our bodies were happier without cheese. And peanut butter could well be an occasional instead of a common addition.

Our own dried apples took on an added importance, as did gathered wild fruits. We saw and learned of food possibilities in the woods and fields around us that we had never bothered with before. And my own encouragement to others to Grow What You Eat and Eat What You Grow took on a deeper and more personal meaning.

It was a good experience and one that hasn't ended. We continue to spend less money on outside food. And the garden reflects even more what we actually eat. More space is planted in the basic foods (corn, beans, peas, potatoes, onions, squash) and less in the minor ones.

We are more aware of the food we do buy. Which foods are more important, which are not necessary but nice, which are simply a treat. And we know, if we should choose or need to get all of our food from our garden and the surrounding wood and field, we could do so. And not suffer a bit.

Go ahead and do it yourself. You'll view your food a lot differently I think. In a positive way. And you don't have to be growing all your own to get involved in this experiment. Expand the realm just a little, which would make sense given a situation where your community, not just yourselves, had to depend on the local area for food. Eat for two months only what is grown and gathered in the immediate area. Or even just one month. But give it enough time to get into it and live it. To adapt, to come up with your own unique solutions for preparing what food you have. To change habits and beliefs. But don't forget to have fun doing it."

"We could do that, JJ. We can go to the farmer's market Saturday. And we can start right now with what is in our garden patch. And we can pick raspberries and blueberries and blackberries and ..."

"Now hold on there, CindyLou. When are we supposed to do all that? You keep telling me we have to get the cabin done because it's going to be snowing soon. And that we have to get our firewood in. And get our garden done. A body can't do ten things at once you know. I think we have enough going right now. We can always do it later. Once we have the cabin done and the garden and the woodshed and the sauna and everything else."

"JJ is probably right, CindyLou. To live entirely off the land takes time, no doubt about it. You might want to save this experience for when you're more settled. But that doesn't mean you can't eat close to the Earth, and this area, right now. Shopping at the farmer's market and buying direct from the growers is a great idea. Just for everyday eating."

"You may be right. I do remember now that bears also like raspberries and blueberries and all the other berries. Maybe we do not have time to venture out into the woods a great deal right now. But I will be back to help you to finish the pickles tomorrow. I can not wait to eat one. Though I do hope they look more like pickles when they are done."

"They will. But they won't be ready to eat for a month or so. You have to give the spices and juice time to do their thing with the cucumbers. I'll write down the pickle recipe for you before I forget."

"Thank you. I guess I will wait for the pickles, though I did not know I had to wait *that* long. And our own Eat What You Grow experience will have to come later, too. Just eating what others grow is maybe enough for us right now when I think of it. I am still not sure how to fix a meal out of much of what is growing in the garden. But we are learning. And we *will* visit the farmer's market. Let us go, JJ. We have those books to read. And I want to look up how to grow Pickling Spice and salt. We have to be able to make our own pickles. See you later, Sue."

"OK, CindyLou. JJ. But about the salt . . . Well, we'll see you tomorrow."

―――――――――――――― ~ ~ ~ ――――――――――――――

PICKLES

Harvest, weigh, then scrub cucumbers (and/or squash, immature watermelon or muskmelon, onions, peppers, etc.)

Slice cucumbers, and other vegetables, to a uniform size.

Cover with a brine made of $\frac{1}{3}$ cup pickling salt to one quart of water [1:12 ratio].

Soak for about 24 hours.

Drain cucumber slices, cover with boiling water then drain again.

Put one teaspoon mixed pickling spices and one dill head into each pint jar.

Pack jars with cucumbers.

Add boiling pickling juice to within ½" of top.

Process in a boiling water bath for 10 minutes.

Pickling juice:

For every pound of cucumbers mix:

1.2 cups vinegar (apple cider or good homemade apple/wine vinegar)

0.6 cups maple syrup (or honey or brown sugar)
[a ratio of 2:1 vinegar to sweetening]

Yield is approximately 1 ½ pints per pound of cucumbers.

~ ~ ~ *twelve* ~ ~ ~

Power for the People
The Sun and Solar of Electricity

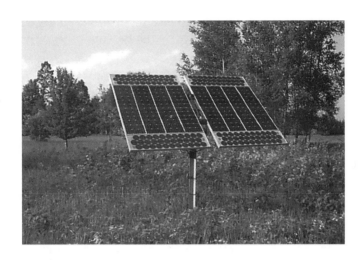

――――――――――― ~ ~ ~ ―――――――――――

"Hi there JJ. What are you working on so vigorously in that garbage can? Looks like quite an aerobic way to take a bath."

"*I'm* not taking a bath, these clothes are. At least that's the way it's supposed to work. I casually mentioned to CindyLou that she'd save a whole lot of money if she'd stop going so much to the laundromat. All I meant was that she didn't have to have things so darned clean all the time. A little dirt never hurt anyone. Well, one thing led to another and this is her idea of a low powered homestead washing machine. Seemed a lot easier when we were talking about it last night. Especially when I didn't figure on me doing the powering. Anyway, I think these things are washed enough. Been working on them with this plunger for probably an hour."

"You have not, JJ. It has only been a few minutes. Just because we are living close to the Earth does not mean we have to wear Her all the time. Hello Sue. How do you like our new washing machine?"

"Quite inventive, CindyLou, seems like it would work just fine. You don't have to worry about many parts wearing out."

"How about me? This body's not getting any younger you know and it can wear out just fine let me tell you. And if I keep this up I'm going to be so wrinkled you won't be able to smooth me out ever!"

"Keep plunging, JJ, so we can get the next load going."

"*Next* load? You said you only had a *little* laundry to do! How many loads are you planning to hit me with anyway?"

"Only three. And it is not mine, it is ours. Here is a basket to wring the clothes into when you are done. It is such a beautiful day to be doing laundry is it not? This is much better than going to the laundromat. I am glad JJ thought of it."

"This is not exactly what I had in mind when I suggested you do the laundry at home, CindyLou! There, that is enough. Come help wring these things out. Boy are they heavy when they're wet. A body could get a hernia wringing out laundry."

"Have room for another set of hands? I'll wring a few for you. Makes you appreciate the old hand wringers doesn't it. We had one for awhile, and we did enough laundry by hand to appreciate the washing machine."

"I would also like to have a washing machine like yours, with the wringers on the side. But we have to have power to run that. And I do not like the sound of that gasoline engine. JJ has been reading about solar electrical systems. But I think it will be a very long time before we have our own. JJ insists he has to do more research. And I do not understand it all that much myself."

"Well now, I'm thinking maybe it won't take as much research as I thought. Maybe we can get going on our own electrical system real soon. *Reeeealll* soon."

"I agree that you need to do your research well before installing a solar electric system, or wind or hydro. But it doesn't have to be a complicated system, especially to begin with. You can always add on later, as you have more time and money. That's how we did it. It's nice to have a little electrical power help now and then. And the gasoline generator is a good incentive for getting your own solar system hooked up. As is doing laundry by hand and dealing with kerosene lamps. It makes a few other homestead chores easier too."

"I would like to have the electricity from the sun but I think those solar panels will not look good on the front of the cabin. JJ says I can not have them on the back but I do not see why not. It is light on the back as well as on the front. I think he is just being difficult."

"I am not, CindyLou. I told you, you have to have sun shining on your panels to make them work and the sun does not shine on the north side of our cabin. You can't have them on the back of the cabin."

"Well, if they have to be on the front then get them in brown. I do not think the blue will go well. Not that your panels are not very nice looking, Sue. But I prefer brown."

"I don't think you have that choice, CindyLou. They don't come in different colors. Most of them are blue; it has to do with what they are made of. Except for some older used units which were brownish. But when you think of the alternatives the color doesn't seem so bad. I do know what you mean though. The usual romantic vision of a cabin off in the woods doesn't include large glass panels across the front! We had our solar electric panels near and on the house for many years before we moved them out into the field for better sun. And you know, now I miss seeing them on the house! They grow on you. Especially when you start using that sun generated electricity. I'm still not fond of the look of the large solar heating units on the front of our house. But when that heat pours in when it's 10° outside and we can bank the fire I love them!"

"I suppose I could get used to the look. Let me get the next load of clothes in our garbage washer so you can get to plunging them, JJ. I do want to get them hung out while it is still sunny."

"I say all this cleanliness is hurting me a whole lot more than a little dirt ever did. So how soon could we get our solar system up and running, Sue? I've been pouring over all those *Home Power* magazines Steve lent me but there are so many ideas I hardly know where to start. There's some pretty nice systems in there. Hey, hold on there, CindyLou, you put much more in there and I'll have to rig up a weight to help me! You know, the laundromat really didn't cost *that* much money now that I think of it."

"You said you wanted to build muscles, JJ. I am just helping you out. Now do not slop so much water out. We do not have that much rainwater saved up and we have to rinse yet. The laundromat does cost too much money, I have come to agree with you on that. You had a very good idea to do our laundry at home. You do come up with some good ideas sometimes, JJ."

"Yeah, thanks. You sure you don't want to build up some muscles yourself?"

"My muscles are already doing very well. And I am not tall enough to do a good job of plunging. We must share the chores as they fit us. I have also looked at some of the *Home Power* magazines, Sue, and I do not see where we can afford many of those systems. And some of them are talking about running many appliances. We have need of just a little bit of electricity. Except in the winter when it will need to make a lot of heat."

"First of all, the panels which produce electricity and the panels which produce heat are two quite different things. And generally it isn't at all efficient to produce heat via electricity. Except for short spurts such as a hair dryer. How about if we talk about the electricity producing solar panels first. Then we can look into solar heaters later."

"Well, electricity is what I'm interested in right at this minute, not heat! There's quite a lack of the first and more than enough of the later around here right now."

"OK. I'll give you an idea for a basic photovoltaic (i.e. solar panel) system. How much usable electricity you can get from this setup will depend on where you are located, both geographically and site specifically. And what Mother Nature is doing with the weather. You need sun to shine on your solar panels in order for them to convert that sun energy into electricity for you to use. So one of the first things to consider is your site.

Look at where you think you want to put your solar panels. Consider a number of different places. Track the sun's pattern and draw sun charts. Few sites are 100% great. You will probably have to figure and think and compromise. But that's half the fun!

There is a nifty devise called the Solar Pathfinder that can help you chart your site for direct sun for every month of the year. It's an easy way to choose the best site for your own array. And the folks who make and sell these are homesteaders too. You can buy one or you can check with a nearby alternative energy dealer. Most offer a site survey service. When you're done with the laundry we can bring ours over and check out your site.

You can also make your own sun charting device. There are instructions in *The Passive Solar Energy Book* by Edward Mazria. Or you can just observe your site over a year's time. You want to put your solar array where it is going to get the most sun when you need it most. That is generally in the winter months when the days are short, the sun low, and you're usually using more power, such as for lighting. Most systems use a gasoline generator to help get them through the worst of the winter months, which for us is November and December. Not only short days but very cloudy ones.

Usually the closer to your home, and the batteries, you can place your panels the better. The farther away you get the more expensive it is going to be, for larger wire or more equipment. Though that may be the better choice for more and better sun.

This is an overview of a fairly simple system so we won't get into all of the equipment you can use in a solar electric system. Pages can be, and are, written about all that. And you've been looking at one of the best sources of information around, *Home Power* magazine.

This is an area where I'm all for JJ's idea of research and more research. Because it's not a armchair project, it's definitely for those who want to get involved. You need to know what you're doing, or get someone who knows what they are doing, in order to install your system safely. Don't get me wrong, it can, and is, a do it yourself project too.

I think you won't have any trouble doing your own system though, JJ. You have enough basic understanding of electricity. And you're willing to learn. That's what will make it affordable when you're on a tight budget.

The total cost is quite variable depending on what equipment you choose, how large of a system you want, where your panels and batteries are located, and how much of the work you choose to do yourself. But a small system will run maybe $750 to $3500.

A basic Homestead System would include a number of pieces of equipment; photovoltaic (PV) panels and rack, batteries and battery box, meters, disconnects and fuses, wire, and possibly a charge controller and an inverter. Let's look at them one at a time.

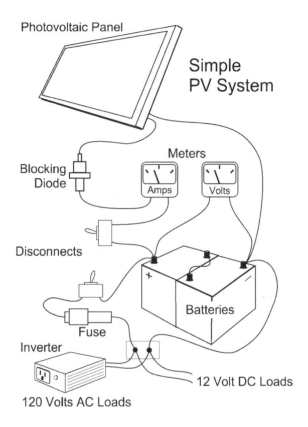

Photovoltaic Panel

Simple
PV System

Meters

Blocking
Diode

Amps

Volts

Disconnects

+

Batteries

−

Fuse

Inverter

12 Volt DC Loads

120 Volts AC Loads

The heart of the system is the **batteries**. If you want
to know how long someone has been involved in
alternative energy just ask them how they started. You'll
probably get a variation on the tale of "Well, I had an
old car battery and wired it up to run a few tail light
bulbs for lights and a radio. When it ran down I'd jury
rig it up to my truck and charge it up while I ran into
town - or - I'd haul my battery over to a neighbor's house
to charge it up." And since the battery was never
intended for that kind of use it would soon die and
another old battery put into service.

We started that way, though we used the 12 volt
marine deep cycle battery. They last longer than a
regular car battery. But they're still not the best, no

matter what their name or advertising suggests. Some lucky (smart) people started right off with a good home system battery - the golf cart battery. Available, recyclable, relatively inexpensive. With good care they can last 5 to 8 years. And they can limp along much longer than that when necessary.

The golf cart battery is a 6 volt battery so you need two, or sets of two, for your 12 volt system. One set will give you a 200 amp hour battery bank for about $150. To help the batteries to a good, long life plan on not using more than 50-60% of that capacity. So with one set of batteries you can figure on about 110 amp hours of usable energy.

You need a good place to put your batteries. I know, many people, us included, got by just fine sticking them under the kitchen counter or wherever. But batteries can be dangerous things. Just because they're common doesn't mean you shouldn't recognize, and respect, their power. They *can* explode. Now this doesn't happen often. It's not something to get carried away worrying about. Just use common sense. A simple enclosed box with a vent to the outside (batteries give off explodable gasses when charging) will do just fine. Keep in mind though that you want to easily be able to access the batteries for checking the water levels and adding distilled water when needed. And keep a good supply of baking soda on hand near by just in case a spill should occur. If you take good care of your batteries they'll give you years of good service.

Now the showy part - the **photovoltaic (solar) panels**. They come in many brands and sizes, new and used, big and small. They each have their pluses and minuses but mostly one is as good as another. The regular new panels (roughly 50-90 watt units) come with a guarantee of at least ten years, some twenty. No one knows how long they will last, they simply haven't been around that long. Used panels can be a good option. They usually don't give as much power for the size as new ones do, but they still produce the power. The panels we bought in 1981 are producing just as much power today as they did then. As I said, there are many options. We'll start with a 64 watt new panel which will cost about $450 and puts out about 3.8 amps of power in full sun.

You'll need a good **rack** for your panel. They can be purchased new or you can make one yourself. You want it to be sturdy but it doesn't have to be expensive or fancy. It can mount on your roof or on a pole, or between two poles. The easiest way is to mount the rack at one angle which would be a compromise between the best angles of each month of the year. Usually the choice is to aim it for optimal spring/fall sun. Or you can make the rack with a simple adjustment that you would set two or four times a year to aim the panels more directly at the moving sun. Or you can go all the way and make it fully adjustable to follow the sun all year. In practical use though this isn't necessary. Usually a twice a year adjustment will make enough of a difference to be worth while but not be too much of a hassle.

The **control panel** for a system such as this can be fairly simple. A Schottky blocking diode between the panel and the batteries will prevent bleeding of your battery power back into the panels (and into the night) when they are not charging. Cost is about $5. You should also have a disconnect, such as a 25 amp SPST switch ($7) between the panel and battery. Another disconnect should be installed between the battery and all loads, along with an automotive type in-line fuse.

You will also want a minimum **metering** of at least an inexpensive digital multi meter which you can get for $30-40. You need to keep track of your battery voltage to monitor how full, or empty, your batteries are (power wise). The meter also comes in handy for various other checking and monitoring chores. Additional in-line meters are nice to have. I'd recommend one or two if you can swing it. An Emico analog amp meter (to monitor your 12 volt use) is about $24. And simple volt meters (analog, digital or LED) can be purchased for $18 to $50. Put your meters where they can be easily seen. Not only for regular monitoring purposes but for the meter reading mania that comes over folks with home power systems.

Then there is the **wire**. An essential, important, and sometimes overlooked part of the system. Twelve volt systems require larger wire than 110 volt. You can find wire charts in many alternative energy catalogues and articles in *Home Power* to help you decide the size and wire to use. The main considerations are distance from one component to another, system voltage, and amount of amperage running through the wire. Whoever you buy your components from will be able to help you with this.

For this system we're going to plan on the panels being within forty feet of the batteries. We'll size the wire to handle our anticipated future system of two modules (total of 7.4 amps). So, 80 feet of #8 USE wire at about $30 should do us. This wire will connect the panels to the battery.

Battery interconnects to connect the two 6 volt batteries together to form a 12 volt battery can be 2 gauge, two cables for $12. For our current system lighter cables would do but for the price we might as well go with the heavier wire and be prepared for future upgrades. You will also need appropriately sized wire from the battery to the loads, such as lights and radio.

Now, there isn't much chance of overcharging your 220 amp hour battery bank with one 3.8 amp module if you're using the power now and then, and monitoring it regularly. But if you'll be leaving the system for some time without using power or monitoring it then you'll want a **charge controller** to prevent the PV module from overcharging your battery. This gets more important as you add panels and additional charging capacity. An 8 amp charge controller is around $60.

So what can you run with one or two PV modules? That depends on your **use**. You can get a whole lot more hours of lighting if you use efficient lights and turn them off when not in use than if you have incandescent bulbs on all over the place. There is a very direct cause and affect here. A closet or pantry light that gets turned on and off often but runs briefly can be a low wattage incandescent. A light that is on for a hour or more at a time should be a fluorescent or compact fluorescent. Lights needed for only minimal illumination such as an entryway or hallway can be small amperage bulbs.

Don't use any more light than you need and put it where you are going to use it. And, of course, turn it off when it's not needed. This goes for any and all electronic appliances. **Conservation** is *the* most important part of the alternative energy system (or, for that matter, any energy system). If you don't use it, it doesn't have to be produced or transported in the first place. So spend some time looking closely at your electrical use, and how you can reduce it.

To give you an idea, Steve and I lived comfortably for ten years on two panels (older 2.2 amp models) running lights, radio/tape player, computer, printers and small tools. For many years we simply reduced our use in the short-sun days of winter and went back to candles and kerosene lamps for lighting. In the summer we had more than enough power. At some point we purchased a gasoline generator to run the washer and the larger power tools.

The generator also anemically served as a backup battery charger in the winter. Until Steve made his own, much better, gasoline battery charger using an old car alternator and the engine from our tiller.

Over the more recent years our power usage has gone up. Especially by the computers and related equipment since that is a good part of our livelihood. We also wanted to get away from using the gasoline generator. So little by little, piece of piece we added to our system. Now we have twelve solar panels, all of the older types, or a total capacity of about 28 amps with two battery packs of six golf cart batteries each. We also have two inverters to run the 120 volt equipment. Now we only have to run the gasoline generator occasionally in a particularly cloudy November or December. Some years it doesn't run at all. Our goal is to dismantle it for good. Along with lights and computer equipment we run power tools, washing machine, flour grinder, and occasionally in the summer, an electric lawn mower to knock down a path or two around our place.

An **inverter** is probably one of the first options you will want to consider for your system. So far we've been talking about a simple 12 volt system. That means all of your lighting, tools, appliances that you want to run will need to be 12 volt. Unfortunately, 12 volt lights are more expensive than 110 volt ones. And the tools and appliances are often inferior as well as more expensive. But 12 volt systems are simple and straightforward, and we run all of our lights, radio, washer, flour grinder, and some power tools on 12 volts. However, to run 110 volt appliances and lighting you will need an inverter to change the 12 volts from your batteries to the 110 volts they need to run. The down side is that it is one more piece of equipment to buy and be dependent upon. But today's inverters are very reliable and efficient so it is a reasonable option.

You can still get the square wave inverters like our old TrippLite which used to run, with a quite annoying buzz, our sewing machine and small power drill. And for certain applications they are an appropriate, inexpensive solution. But for most home power use you will want a modified square wave inverter. This will run most things just fine. There are many models and brands and they all have their pros and cons and loyal users. You can check out the *Home Power* "Things That Work" and other articles and talk with your supplier for recommendations. Intended use and price will probably dictate which one you will choose. They come in all sizes and many shapes.

You may also want to consider a true sine wave inverter. Especially for special applications such as a laser printer which require that "cleaner" power. They also run some other appliances better than a modified square wave inverter does. We have a 1300 watt modified square wave inverter which powers the computers, bubble jet printer, power tools, sewing machine, vacuum cleaner. And a 500 watt sine wave inverter for the laser printer and scanner. Inverter technology is changing all the time with larger, more efficient models. Look closely at your needs and get the model to fit. No use spending more money than necessary. For years we made good use of a small 100 watt modified sine wave inverter to run the computer.

Remember, however, an inverter does not produce more power, it just helps you use it. No use buying a 2500 watt inverter for a 1200 watt sized system. But with the efficiency of today's inverters many people are choosing to run their entire place through an inverter. That way you can use more commonly available 110 volt wiring and lighting and appliances. This is certainly an option to consider.

But we're talking small and inexpensive here. Your basic 12 volt system for a small cabin/home. If there is a particular 110 volt appliance or tool that you want to run, without running the gasoline generator, you can buy an inverter just for that, say to occasionally run a power drill or a computer. Small 150 to 250 watt inverters cost about $100 - $150.

Another important aspect of your system is **labeling**. Do a lot of it. Make it easy, make it readable. Do it for someone who has no idea what your system is all about. Assume that someday someone else will need to walk into your home and be able to figure out your electrical system. Of course you will want to make sure you understand the system yourselves, as well as does anyone who will be staying in your home. You are the power company. You want to make your system a safe and workable one.

So what is the total cost today for a one module Homestead Solar Electric System? Assuming you are doing the work yourself maybe $750 to $1000. Where to get the equipment and more information? Well, you already have one of the best sources for information in the *Home Power* magazines. Even if you weren't going to be doing the work yourself I'd recommend reading as much as you can about the subject. And that's the best place I know to get information on suppliers. Send for catalogues, talk with the dealers. You want to find a dealer you are comfortable working with whether mail order or in person, then work with them. When you add up prices for the same components including shipping and handling you'll find little difference in the total price of a system among most dealers. Service is by far the more important. You'll want to deal with someone who can help you, within reason of course, and someone you can trust. I think attitude is more important than price.

For a local dealer you can look through the *Home Power* magazine to see if there is one listed who is near you. But for much of the country you'll probably have to do some asking and checking around. Many alternative energy businesses are small and regional. They aren't going to have big budgets for advertising and probably get almost all of their business word of mouth.

Another good source for A.E. information are the many regional energy fairs being held throughout the country. The largest is the annual Midwest Renewable Energy Fair in Amherst, Wisconsin, which isn't all that far from here. Held the Summer Solstice weekend this is a three day affair packed to overflowing with workshops, displays, and events (and people!). There are many other good regional fairs too.

I think the most important thing is to actually get started. Don't wait until you can afford your dream system. Start with what you can. Even those who are hooked into the grid can put one small circuit on their own PV system. They will find themselves with lights and radio when the rest of the neighborhood is blacked out with a power outage! And those who are running from a gasoline or propane generator really have a treat with a battery system which allows them to turn on a light or watch TV without that generator guzzling fuel and roaring away in the background. PV power is not costless, but it is quiet!

And, it will let you use your muscles for something else other than plunging clothes, JJ! Not that I think this is a bad way to use your muscles or your time. But come winter you might appreciate an electrically powered wringer washing machine. Looks like the laundry is done. Nice job, JJ."

"Whhheewwww! Yep, a solar electric system is next, CindyLou, no matter where we put it. Just think, reading all night without being bugged about using up the candles. And a washing machine and electric dryer and corn popper and coffee maker and microwave and waffle maker and large screen TV and stereo in every room and lights and whirlpool bath and more lights and . . ."

"And moving out. Grab that basket, JJ, let us get these clothes hung up before nightfall."

"You want me to move? Gee, no respect for this old body. You never even give it a chance to set up. A body's got to set up once in awhile in order to creak properly you know. And I was just kidding. I know we can't have all that stuff. Or want to, or want to. Unruffle there. Though hot waffles in the morning . . . OK, OK. Sure glad you didn't run those clothes lines very far away. Even though I'm likely to hang myself some night wandering around in the dark."

"Use a forked stick to hold the lines up when you're not using them, or when you're done hanging up the clothes. That way they can be low enough for CindyLou yet high enough to get under. As far as what you can run with your solar electric system, JJ, it's only going to work if you're conservative you know. I think you better think on the order of a few lights and radio and an occasional larger load. You'll have to sit down and decide what is important to you and what will fit your budget. But I think you'll be happier popping your popcorn by hand in a pan, and pancakes are just as good as waffles. They're both pretty easy, and you can make them using what power you already have, called human muscles. It's inexpensive, and better looking, too!"

"Well now, I have to admit this old body *is* getting some pretty good looking muscles, if I do say so myself. Not bad, not bad at all. And I must say my pancakes are something to wake up to, so it's been said."

"As long as you keep hanging up laundry, JJ, you can say whatever you want. And since you mentioned it, I would be very happy to wake up to pancakes on the table tomorrow morning. Yes, that is a very good idea you have there. I am looking forward to that very much. Do you want to come over for pancakes early tomorrow morning, Sue?"

"Now, CindyLou, that's not quite what I said! I'm sure I didn't . . ."

"Thank you, CindyLou, but I think I will pass, it sounds like a private breakfast. And I had better be getting back. Let us know if you have any questions on your solar system. See you two later."

"Yep. Now, CindyLou, it's not that I don't want to make you pancakes but getting up first early in the morning, well, you know this old body needs its sleep, it's not getting any younger you know, and . . ."

~ ~ ~ *thirteen* ~ ~ ~

Age Old Meets New Age at the Solar Food Dryer

~ ~ ~

"That's quite a pile of apples you're cutting up there, Sue. You aren't by chance planning to make a few dozen pies or something are you? I sure could figure out what to do with them if you find you have too many."

"JJ, mind your manners! We did not come over here to beg. But I must say I do wonder myself what you are doing. You do not have your cook stove going so can not be planning to can applesauce."

"Sorry, JJ, no pies. I'm afraid I'm not much of a pie maker. I'd rather just eat the fruit straight. And no, I'm not making applesauce, CindyLou. At least not now. I'm making dried apples. Or cutting up apples to dry them. It's easier and faster than canning. They make a great snack food or we can cook them up into sauce. Dried apples make great traveling food too, whether hiking, biking, or in the car."

"Looks like you're going to have enough dried apples to feed the entire neighborhood! Looks like a lot of work too. You know, maybe we better come back later, CindyLou, when Sue's not so busy. Don't want to bother her you know."

"Drying apples would be a very good idea for us, JJ. We can collect enough apples just walking up and down the road to have a very good many to cut up and dry. And it will not cost anything. Meantime, we can help Sue so we know what to do with our own. And find out what she does with them. If you do not mind of course. I do not want to bother you overmuch."

"Of course I don't mind, CindyLou. I'd appreciate the company. Here's another cutting board and knife. I'm cutting them into cubes of about $1/2$". They need to be fairly uniform in size so they will all dry about the same time. If you'd like you can wash another basin of apples, JJ."

"Wash, wash. Seems like whenever I'm not building I'm washing. You could be washing off very important and healthy bacterias you know."

"Could be. But I think I'd like the apples washed anyway. Personal preference."

"And you're just throwing those skins out anyway. Why bother to wash them? And I just know this is all leading somewhere that I probably don't want to know about. Nothing's ever just a simple project."

"Could be. We'll come to that in a minute I think. But I'm not throwing the skins out, JJ. They're going into that clean bucket, along with the cores, to make vinegar. I don't always peel the apples before cutting them up to dry. They do just fine with the peels on. I'm just peeling this batch for comparison, see if I like them better peeled. I'll probably decide the unpeeled dried apples are better since that's easiest. Though for canned applesauce I peel them, it makes a better product. And peeling gives me more material for vinegar making."

"What are you going to do with all those apple cubes? If you just lay them out in the sun to dry the birds will eat them. And who knows who might crawl across them."

"Well, I'm not too concerned about that, even if I did just put them out in the sun to dry. It's a system that has worked for hundreds of years so I'm sure it would work fine now as well. I read in a book published in 1782, *Letters From An American Farmer,* by J. Michel Guillaume St John de Crevecoeur, a good description of drying apples. He said strong crotched poles were planted in the ground in a place where the cattle couldn't come. Horizontal poles were set into these and boards laid close across. Then the apples were thinly spread over it all. He said they were soon covered with bees and wasps and sucking flies and that this accelerated the operation of drying. Now and then the apples were turned. And he was talking about twenty baskets of cut up apples! At night they were covered with blankets. The skins and cores were dried with equal care and used to brew a special beer, both for drinking and for the yeast which was formed and used in the baking of bread."

"Twenty baskets full? They must have eaten a *lot* of apples and had permanently wrinkled hands from washing them all, which I am on my way to having myself. But now I like the sound of brewing that special beer. Sounds a better use for those peels and cores than vinegar. Think of the bouquet that a brew such as that would elicit!"

"You had better think more about all that water you are dripping on the floor, JJ. You are making a mess of the rug."

"Oh sorry. I got carried away with the vision."

"Not to worry, JJ, it'll dry. Unfortunately, I don't know how to make the brew he mentioned, he didn't include a recipe or any other words about it. I'm more of a wine maker myself, it's easier. And the vinegar I make is important, too."

"I would like to know how to make vinegar and wine. But right now are you going to build a stand outside to put your apples on? It does not seem as if they would dry well. Even when it is not raining outside it does not feel very dry."

"That's true, CindyLou. In the fall and winter when the wood heating stove is going I dry things in the racks over the stoves. There is one over both the heating stove and the cook stove. They are simple racks which hold frames of screening. Later apples are dried there, as well as herbs and vegetables. We have dried a lot of food over the years with this simple system. And the racks and screens are easy to make.

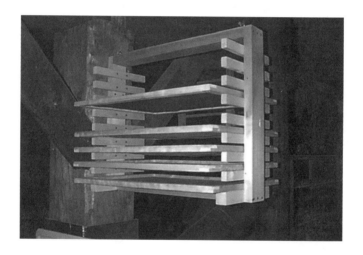

But in the summer and early fall we have another solution for our humid Midwest. A solar food dryer. We can dry more food at once this way, and faster too. As long as the sun is shining that is. I think we have enough apples cut up now to fill several screens. Then we can take them out and I'll show you the dryer."

"I knew there was going to be more to this. You always bring it around to building something!"

"Of course, JJ, I wouldn't want to disappoint you!

But they're good projects, both the indoor racks and the outdoor solar dryer. And they are ones you'll come to appreciate. Drying is so much easier and faster than canning. And no firewood to have to collect for it. Once the dryer is made you just use it. I think these are projects you'll like.

We'll just spread our apple pieces out on the screens. The rack above the cook stove was built to hold these particular screens. It's handy for holding the filled screens while you fill another. And when it clouds up or rains I bring the filled screens from the solar dryer outside to the inside. It's usually cool enough when we have that kind of weather that we'll start up the cook stove. That helps the food dry some while it's waiting for a sunny day to go back out in the solar dryer. Now let's get these apples out into the dryer.

You have to keep an eye to the weather when you're planning to dry a lot of anything and some items are finickier than others. Ideally, they would go in the dryer and stay until well dried. But often you have to work around a sudden cloudy day or a busy schedule. It's quite important to get them on their way to drying before they have to be held over due to damp weather. So I usually process things for the dryer first thing in the morning on what looks to be a nice sunny day. That way they'll get a good start on the drying even if it should turn cloudy the next day. Or if you think it will be sunny the next day you can prepare your food the night before so you can get it out in the dryer as soon as the dew is off in the morning. This is good for green corn which takes a while to prepare.

Things like herbs and greens aren't much of a problem, they dry fast and can be done in a day. Apples don't take too long, nor do chopped onions or celery or broccoli. But something like green corn or strawberries take longer. And tomatoes are particularly full of moisture and take a good long dry spell to dry them without molding. Most vegetables and fruits are pretty easy though. I usually don't dry herbs in the dryer because it's easier to just hang them up inside. That way I don't have to worry about over-drying them.

The books I've read about drying recommend blanching almost everything first. And I do steam blanch my green corn, and snap beans, and carrots. But I've found many things dry just fine without that blanching step. Apples certainly don't need it. You should experiment yourself to see what you prefer.

There are many ideas and designs for solar dryers. They certainly aren't a new invention. But most of those designs are for areas with generally dry weather. Not something that can be said about the Midwest. The idea for our dryer came from Larisa Walk and Bob Dahse of Minnesota. I first saw it written up in Issue #29 of *Home Power*. That article also had a dryer for the dryer climates. We were fortunate after that to meet and share gardening and homesteading ideas with Larisa and Bob at the Midwest Renewable Energy Fair in Amherst, Wisconsin. It was after we talked with them and fellow drying enthusiast Kathleen Jarschke-Schultze (whom you'll find often in the pages of *Home Power*) that we finally decided to make our own dryer. As with so many projects it turned out to be something I wish we had done many, many years ago.

Larisa has this to say about how the idea for their dryer came about:

"For years I tried about every solar dryer design imaginable. The only common factor in all those attempts was their very limited usefulness here in the humid upper Midwest. None of them could reliably turn food into a non-moldy finished product, unlike the many successful electric models I had built for myself and friends. Some didn't work at all if not tracked periodically during the day. It was with this background that the 'idea light' came on in the head.

"One day I needed to dry a bunch of greens and the current solar dryer was full (a couple of handfuls was all it could handle). I had an old window screen lying around and a corrugated metal roof built over our old trailer house. Using a ladder to get to the roof, I put the screen down first and put the food on it. I wanted to keep the sun off the food itself so I covered it with a piece of black cloth. Then to keep everything from blowing away, or being bothered by flies, I covered it with the storm window that was lying around with the screen.

"Later that afternoon I thought I'd see how it was doing. The greens in the old dryer were still quite limp when I crawled up the ladder to take a look at the stuff on the roof. Much to my surprise, the roof-top greens were crispy dry! It looked like I had finally stumbled on something that worked. I tried several other foods on the roof before I was convinced enough of the design to build a unit at ground level for easier access.

I found through experimenting that the primary ingredients for this idea were: 1) Corrugated, galvanized metal roofing, 2) Screen, 3) Black cloth (or black metal), 4) Glazing, 5) Slope. The sun shines through the clear glazing onto the black cloth, heating up the air space under the glazing. The corrugated metal provides air spaces under the screen for the warm, moisture laden air to move. The air moves passively upward along the slope, carrying away the moisture from under the trays of food. The galvanized metal also reflects heat back onto the food. This combination really gets the job done."

We used Larisa and Bob's basic idea along with what materials we had on hand to build our dryer. We designed it around our already existing cold frame windows. This works great since by the time I'm ready for the dryer I'm done with the cold frames. Each set of two windows is hinged in the middle with loose pin hinges so they come apart easily. And each window has a simple handle on the outside edge, which is useful both on the dryer and on the cold frames. We made the windows ourselves and the finish is a simple raw linseed oil/turpentine mixture, 2:1. If you use old windows you'll want to scrape any paint off as that isn't something you want falling into your food, or into your garden.

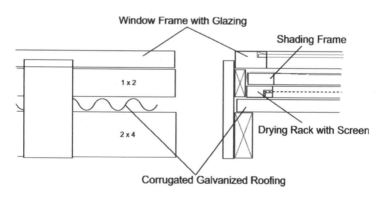

Our dryer frame was built with what wood we had, which happened to be poplar. Not the most durable wood. But it is well oiled and is stored out of the weather when not in use. I'm sure it will give us many years of service. We use regular saw horses for the legs since we have them and they're easily stored when not in use under the dryer or for some other project. Blocks of wood elevate the top of the drier to the right angle to aim at the sun. You could also put simple legs on your dryer, of course, which is what Larisa and Bob do.

The galvanized roofing we bought. Though that is something that in many parts of the country you can probably get a used piece of if you keep your eyes open. But a new piece is not that expensive either. Small boards on the bottom and top, and boards on the sides, give the windows something to rest on above the screens. They have to be tall enough to give enough room for the food drying screens and the shade screens. The bottom blocks are stepped to prevent the windows from sliding off. Make your rack large enough that the you can easily get the screens in and out.

To keep the screens of food up off the metal roofing we lay rounded strips of wood under them. You could also make your screens with a thick enough frame that the screen wouldn't droop down on the metal. This isn't a problem with a sturdy stainless steel screening material. But fiberglass screening does droop with age and use. Larisa's early screens were made with hardware cloth on which she laid the fiberglass screening. For many years I used fiberglass screening stretched in a wooden frame made in a similar manner to regular window screens. And while that worked OK, it did, as I said, get droopier the more it was used. And it's not that easy to clean. It is readily available though and inexpensive.

The ideal screen material is a sturdy stainless steel. Not cheap. And not readily available. But I did finally splurge and it is a great upgrade to the system. Abundant Life Seed Foundation sells stainless steel screening in various fineness including a #8 (wires at eighth inch spacings) which would do. Unfortunately they come in one foot square pieces for seed cleaning screens but could be used for food drying screens if you planned your rack around that size. They are $10 each.

Another source for larger pieces is the Cambridge

Wire Cloth Company, PO Box 399, Cambridge MD 21613, 800-638-9560. This is where I got my stainless steel wire cloth (which is how they refer to it). They are a company who has been in business for over 80 years and they serve customers of all sizes. They do have a $100 minimum order but you can always go in with one or more other solar dryer builders and combine your orders. I didn't have any trouble coming up with an order that large for myself when I added up all of my food drying screens, for the outdoor solar dryer and the indoor over-the-wood-stove rack. Cost depends on what you get. And what you can buy at any given time depends on what they have available. But a SS mesh of .025 size wire is common, and thus less expensive, and a good choice for the screens. I purchased a piece 50" wide by 11 ft long of .025 10x10 mesh (10 wires per inch) for $117.50 including shipping. Or about $2.60 a square foot.

It is expensive but it is very nice to have a screen that is easy to clean and is sturdy. Keeps the peas and corn and such from all rolling into one droop. It is especially nice when you're drying things such as cut up strawberries. They *do* stick! In fact, you want to be sure to peal them off and turn them over before they get stuck so well you have to scrub them off.

The wood for the screen frames can be of whatever you have. They don't have to be oiled or anything. Steve made ours with simple butted angle corners held together with glue and strong staples. A ledge was routed out on the inside edge where the screening is stapled. Then a small strip of wood is nailed on over that. It makes a neat, sturdy screen. You could also staple the screen to the bottom of your frame then nail on a thin piece of wood over it. A screen full of wet food can be quite heavy and you don't want your screening coming off and dumping all that hard labor, not to mention good food, on the ground.

The shade/heat absorber can be made in the same

manner. Larisa recommends black polyester material because it doesn't fade as fast as cotton. If you make your frames as above with a routed in ledge (so the material is more or less in the middle of the wood frame) when one side fades after several years of use you can turn it over and use it for many more years before you need to replace the material. You can also drape the material right over the food if you want. But having it in the frame makes it much easier to handle. And cleaner too.

Larisa said they recently discovered that black metal works as well as the cloth and lasts longer. It is what they are using now. They use aluminum flashing painted with BBQ black on both sides. You could also use printers plates or sheet steel. They staple this to the underside of their cover frame which is glazed with Kalwall or corrugated fiber glass greenhouse glazing. The metal could also be stapled to a separate wood frame so it would be removable.

You'll have to experiment with your dryer in your climate and conditions to see how it performs. On clear, hot sunny days I have to block the windows up some to prevent the dryer from overheating. But it's nice to have a closer fit for partly cloudy or somewhat overcast days. You *can* burn your food in there! And from experience I can tell you there isn't much of a call for blackened broccoli. Just as different foods will take different amounts of time to dry, different weather conditions will also affect the drying times.

Once you start drying you probably won't need any encouragement to experiment. A couple of successes I was especially happy with was dried cooked beans or peas (appropriately cooked in a solar oven). Crack them in a steel burr mill after drying and they cook up fast. Also good are cooked grains such as rice. Homestead instant food! Great for camping or last minute meals. Cooked lentils dried made a particularly interesting food as they could even be eaten as is without rehydrating and cooking.

Dried foods aren't going to be the same as canned or frozen. But they don't have to be. They are a good food in their own right. Drying food is a great preservation method; it costs little, is easy, and the stored food takes up relatively little space. A lot less than fresh, canned, or frozen.

You know, when this batch of apples is dry the dryer will be free for your use if you want to put up some of your own. No use letting it sit idle. That will help you get some ideas for how you want to build your own."

"Thank you. We will do that. We will take some days off and go searching for apples. If that Michael Kreever or whatever his name was could do twenty bushels then we should be able to do at least ten. Yes, we will make ourselves enough dried apples to have applesauce every day. Come along, JJ, we had better get to searching for apple trees."

"Ten bushels?! Do you have any idea of how much ten bushels of apples weigh? My back aches just thinking about it. And we'd be eating applesauce for breakfast, lunch, and dinner and all through the night besides. A body can't live on applesauce alone you know. Don't go getting carried away with this thing, CindyLou."

"Actually CindyLou, J. Michel Guillaume St John de Crevecoeur didn't process those apples himself, he called in neighbor women, for which they received a good supper and tea for their labors. I think that might be a lot of apples for just JJ and you. Not that I want to discourage you. But maybe for this first year you might want to go with just a few bushels."

"A *few* bushels? That's hardly less than ten! And you know who's going to have to carry them. Why these old muscles are getting used and abused just thinking about it. Why can't you come up with a project that lets me sit in my easy chair the way this old body is supposed to be doing."

"Come on, JJ, we have much to do. And you do not

have an easy chair. *Ohhhh, we'll find us an orchard, with apples and treeeeeees, to dry for the winter, and sing with the beeeeeees."*

"That doesn't make any sense, CindyLou. You don't dry trees, and what do bees have to do with anything? Hey, slow down. A body can't haul apples if it's all worn out from running you know. In fact, I read just the other day that if you don't take care of your body when you're young, not that I'm all that young any more, but relatively speaking you know, why when you . . ."

~ ~ ~ *fourteen* ~ ~ ~

The Days of Wine and Vinegar

~ ~ ~

"Hi there CindyLou, JJ."

"Eeeeckk!"

"Whooohh!!"

"I'm sorry, I didn't mean to startle you. It's just me."

"Of course, of course, didn't think it was anything, or uh, anyone else. Just didn't see you behind that tree there. Didn't actually scare me you know, just surprised me."

"Well, I do not mind admitting that you scared me,

Sue. I am trying to keep the thoughts of bears out of my mind but they do jump in once in a while. Especially when someone makes a loud noise from the trees. When my heart slows down some I will ask what you are doing."

"All I did was say Hi, CindyLou. You two need to spend more time out in the woods so you can get comfortable with it. It really is one of the friendlier places to be."

"One thing at a time, one thing at a time. I'm just learning to live with the mosquitoes and black flies and wood ticks. I'll wait for the bears for later. Not that *I'm* afraid of them. For CindyLou's sake you know. We'll take them on some other time."

"If you are not afraid of the bears why do I always have to go with you when you want to go walking in the woods?"

"Companionship, CindyLou, companionship. You're always going on about how this is *our* project not yours, so I'm just including you in."

"Well, it's a nice day for a walk in the woods for whatever reason. And you don't have to worry about the bears. Going anywhere in particular?"

"Nope, just needed a break. If it was up to CindyLou we'd be working twenty hours a day. Have to drag her away sometimes for her own good. Say, you're not picking those things are you? If she invites us over for pie, CindyLou, be sure to beg off. I ate some of those little cherries and they're soooooooooo puckery it took me all day to get my mouth back in line."

"JJ, they are not that bad. I tried some and they were edible. If one was very hungry. What are they? And I also wonder why you are picking them. There do seem to be better tasting berries around."

"These are wild black cherries. Quite similar to chokecherries. Most of the cherry trees here are wild black cherries, though we do have a couple of small trees of beautiful red chokecherries at the bottom of the old driveway. But you are probably both right on the taste. There is quite a difference among the trees. Now this one here isn't at all bad. Here, try some. Maybe not your tame black cherry flavor but it's the perfect taste for sampling when you're out walking in the woods."

"I don't know if my mouth can stand being puckered up like that again. You try them, CindyLou, maybe your mouth isn't as susceptible."

"Oh, here, eat them, JJ, they will not kill you. We need to expand our food horizons. You never know when we might have to live entirely from what we can gather from the woods. I would not say that these are what I would want to have for dinner every night, but you are right, Sue, they are not too bad. In fact, they are better than the last ones I tried."

"I don't want to even think about contemplating having to live off of what we could find in the woods to eat. A body would starve soon enough if it had to survive long on these chokecherries. And what a fitting name. Oh all right, hand some over . . . Ptuey. OK, so I admit these cherries here are not *all* that bad. Trouble is you'd have to eat cherries from every tree just to find one that doesn't pucker you up permanently. Not a pretty sight to contemplate."

"I guess I have to agree with you on that, JJ! I think you can unpucker now. But you're wrong about not being able to survive on chokecherries. However, I'm not worried about surviving with these right now. I'm harvesting the cherries to make wine."

"Well, now, here's a project I can get interested in! I don't think even you can come up with something that has to be built just to make wine. And a nice smooth silky draught of clear effervescent liquid with just the slightest touch of sweetness sliding down this old throat would surely soothe the most puckered puss. Ah yes, these choking black cherries do have a different look to them now. Maybe we should stop over tonight to help you test this current year's brew."

"You're welcome to come on over, JJ. But all you'll taste is black cherries boiled in water at this point. And I can't say as my wine comes out clear or effervescent. I think of it more as hardy. Suited to a long, cold winter. And we like it on the sweet side so it may not be to your liking."

"It sounds as if it is to my liking though, Sue. I would like to know how to make wine. It seems a waste not to make something out of all of these beautiful little cherries. I just hope it does not take a great deal of time or researching. Winter is coming and we do not have our firewood in yet."

"You can make chokecherry jam and probably pies too. But they do take a great deal of sweetening so I don't bother. I prefer them for wine. And wine doesn't take much time. You *can* spend more time, and money, and there's a lot of researching for recipes and techniques that can be done, which should catch your attention, JJ. But I make a pretty basic, easy wine. And since it usually comes out quite edible I stick with the same recipe."

"Usually? What happens when it's *not* edible? The taste testing could be real hard on a body."

"If it's not good it would go in the compost pile. That only happened once with some rhubarb wine I made. Out of some fifteen years of making wild black cherry wine I only had one that we didn't care to drink. And it was fine for cooking so it got used.

"I've tried pin cherry, strawberry, apricot, and blackberry wine. They were OK but not as good as the black cherry/chokecherry, so I usually stick to that. As you said, CindyLou, it's nice to be able to use these cherries for something.

"I have also made gooseberry wine a few times and wild plum too, both of which were great. But it's hard to come up with enough gooseberries and we don't have wild plums growing here. Though we did plant some wild plum trees some years back so I hope to have them someday. A couple of the trees are looking good."

"Well, I'm hooked. Let's go make wine. We'll be happy to keep you company while you're making it."

"That'll be fine, JJ. But I don't have enough cherries picked yet."

"Oh, well. I suppose . . ."

"Hand us your extra basket there, Sue, and we will help you pick. Come on, JJ, if you want the wine you have to pick the cherries. This is a very interesting berry basket. It looks like it is made of something which was made to look just like bark!"

"It is bark, CindyLou, poplar bark. It's a berry basket, and quite handy for that. Not too hard to make either. I'll show you how next spring if you want."

"Now there you go, taking something as simple as making wine and turning it into a project. I knew you would do that."

"If you do not like the basket you can jog on back to the tent and get a plastic bucket. We have cherries to pick here. And no one said you had to make a basket. Or drink the wine."

"Now, I didn't say I didn't like it. It's a great basket. I just . . . OK, OK, let's pick cherries. This old throat is getting mighty dry waiting for some good old homemade wine."

"Thanks for helping pick cherries JJ, CindyLou. This should make a good batch. Let's measure them into the basin here. Four quarts, not a bad harvest. We'll wash them and pick out the leaves and stems and stuff that wouldn't add to the wine."

"Even if you squeeze really hard and get every last drop of juice that sure isn't going to make much wine. Seems like a lot of work for just a small glassful or two."

"Oh, it'll make more than that, JJ. If you want to finish washing these last ones I'll get the cook stove going. After they've drained you can put them in that stainless steel stock pot, CindyLou.

The recipe is quite simple:
> 1 quart (1 ½ lbs) chokecherries/black cherries
> 10 cups water
> 3 cups (1 ½ pounds) white sugar
> wine yeast

> additional sugar added later if needed

This is the only good use for white sugar that I've found. I'd rather use maple syrup, but my experiments with that haven't been very good. Maple syrup is a little too strong in its own right. It's worth more trial though.

Simmer the berries and water about thirty minutes in a stainless steel or enameled kettle and add the first batch of sugar. Let it cool to around 95°, then add the wine yeast. I use whatever general wine yeast is available locally. The packets I get are for five gallons of wine and contain five ¼ teaspoons of yeast. So for each gallon of wine, I add ¼ teaspoon yeast.

It takes about two pounds of cherries to make one gallon of wine. We have four quarts of cherries here, which is six pounds divided by two equals three gallons of wine. So we will add ¾ teaspoon wine yeast when the time comes. You don't have to worry about being exact, a little more or less isn't going to hurt.

We need a good large container for the wine because when it starts working it foams up a lot. If I don't need my large stainless steel stock pot for anything else, I leave the wine in there. An enameled canner or glazed crock also works, or a good clean five gallon plastic bucket. Cover with a loose fitting lid or cloth and set it aside to work.

After about six days, taste the wine, adding more sugar if it is not sweet enough. It should be definitely sweet but not cloyingly so. Stir well then strain. I strain first through a colander and then through a clean, wet cloth. The leftover pulp is added to the vinegar bucket which already contains apple peelings and cores. This combination makes a great wine vinegar which I'll tell you about later. We don't want to get our wine and vinegar confused. The wine is poured or siphoned into clean glass gallon jugs. But not too full because you have to leave some room for it to work; it will still be foaming.

Now you want to keep other yeasts out of your wine, especially the vinegar ones. At the same time you have to allow the gasses to escape from the working wine. I now use regular water locks which you can get from a specialty wine/beer store. They're nice because they are reusable. But for many years I used balloons. Find some with a wide sturdy mouth. Just stretch the balloon over the top of the jug. Keep at eye on them though because if the balloons are too small they can be blown right off of the jug. Whatever you use set your jugs aside to work.

By December the wine will be ready. I use a plastic tube to siphon it out of the jugs and into clean bottles with screw caps. Keep the tube up off the bottom out of the scum when siphoning. Homemade wine makes a great gift so you'll want to be sure to make enough. For cleaning out the jugs and bottles small clean stones work great. Just don't get too carried away rattling them around so hard you break the bottle.

As I mentioned, gooseberries and wild plums make a good wine too. The recipe is:

1 quart (1 lb) gooseberries or 1 qt (1 ⅓ lbs) wild (Canada) plums

9 cups water

2 ½ cups sugar

¼ tsp wine yeast

1 ¾ cups sugar added later

The quantities and weights for all are approximates. I pick what I have and use these general proportions, rounding off a lot. Before you add the second batch of sugar taste your wine first. Each year is different and you may want to add less or more. Adjust it to fit your own tastes."

"December? That's a long time to wait for a glass of good homemade wine."

"Anything good is worth waiting for, JJ. Besides, we have to go pick more cherries to get our own wine going. We have an enamel canner to cook it in and we can scrub up one of our best 5 gallon plastic pails. I am very excited about this. We are going to make our own wine, JJ! But one thing I do not understand is why that bucket of apples and cherry pulp will be vinegar and what you are simmering here will be wine. I do want to make vinegar too, but I would rather my wine be wine."

"That's why I add wine yeast to what I want to be wine. That gets the right yeasts in the right spot and while it's working it's not going to invite any other yeasts in. At least that is how I suppose it works. If you left your wine out without a water lock or balloon barrier it would probably turn into wine and then into vinegar. But by putting that barrier on the working wine we block out the vinegar yeasts that are probably in the air.

Vinegar is easy to make and is a natural sideline to the usual homestead kitchen activities in the fall. When I start processing apples for drying or applesauce I start making vinegar. The best tasting is the batch that is a combination of apples and wine pulp. This gets used for pickles and salads. The plain apple vinegar is quite variable. The best (strongest) will probably be used for pickles and the worst (weakest) for cleaning. You don't need all that specialized cleaning stuff that's marketed. Vinegar and baking soda may not be as exciting but they are every bit as effective. Cheaper too, and more pleasant to use.

A good clean plastic bucket, or glazed crock, stainless steel or enameled pot, and good water is all you need. When processing apples the cores and peels go in the bucket. As does any other fruit remains such as blackberry pulp from jam making or the black cherry wine strainings. Warm water is added to cover generously and a cloth is draped over the top. This is set in some out of the way but warmish spot.

The first year I simply left the brew out in the kitchen (covered with a cloth) and it picked up the vinegar yeasts just fine. Thereafter I saved the "mother" which forms to add to the new batches, extra insurance to get your brew going down the right track yeast wise. After the brew has turned into vinegar (in two to six weeks) a pinkish soft slime will form over the surface. This is the mother. You can usually gently lift/skim this off (it rather holds together) into a jar and save it in a cool, dark place for future use.

When your brew is vinegar (taste, and the formation of the mother, will let you know when it's done) strain it into clean jars and bottles. Use containers with plastic lids or put a piece of plastic bag under a metal one. This is true for the mother jar too. Vinegar is an acid and will corrode a metal lid. Store in a cool, dark place. If you don't get a good mother don't worry. If it tastes like vinegar it is probably vinegar. Most homestead kitchens should have enough of whatever it is that gets vinegar going to get a new batch going next year. This is a natural process and it doesn't require any fussing.

So, you're all set for making both wine and vinegar. Especially now that you have a place for things such as that."

"Yes, I would not want to have tried to find room in our tent for jugs and buckets of wine and vinegar. It is nice to be in the cabin and to have the space. Even if we are so very far from being finished with the inside. But when we do have something to clean I want to have vinegar to clean it with. Come along, JJ, we have harvesting to do. We will have to do more applesauce so we have apple peelings and cores for vinegar. And we have many cherries to pick. Wild black cherry wine will be good for Christmas presents next December. I do hope there are enough cherries left."

"I'm sure there are, CindyLou, it's a good cherry year. Here, you can borrow the berry baskets. Do you want a glass of last year's wine before you go for some incentive?"

"Thank you Sue, but we will have the wine later I think. We have much to do. Berries to pick and apples to pick and maybe some blackberries or blueberries for an added touch I think."

"Now, CindyLou, Sue's got the right idea. We had better have a taste of what we're trying to make here so we're sure we're doing what we ought to. Wouldn't want to go to all that work for nothing you know."

"You can sample the wine later, JJ, let us get going. The sun does not stay up all night and we have a great deal to harvest. You were the one who was complaining about there not being enough hours in the day. You can drink wine in the dark but you cannot pick cherries in the dark."

"OK, OK, but while I'm all for some good jugs of wine fermenting in our cabin, CindyLou, remember we do not have to pick every cherry there is. Have to leave some for the birds you know. Can't have the birds starving to death because you wanted more jugs of wine. Not that having many jugs of wine is all bad if you didn't have to pick so many cherries to fill them. And as far as more apples, why we pick any more apples there won't be room for us in the cabin. Not that the vinegar isn't a good idea, I do like those pickles Sue made, but one has to do these things in moderation you know. You get too carried away and we'll be having to drink wine and vinegar just to get into our place. Not that a bit of wine now and then isn't very healthy, but that much vinegar could be a bit hard on the body, and . . ."

~ ~ ~ *fifteen* ~ ~ ~

To Save the Garden Seed

~ ~ ~

"Hi there Sue, what're you doing with all those grocery bags? You're a long way from the grocery store."

"Hi JJ. I suppose you could say I'm going shopping for food, in a long term sense of the word. How are you doing today? What did you do with CindyLou?"

"Oh, she had to stop and pick some of those big balls of fluff you have in your field. Says you told her they were edible but I think you were pulling her leg. Even if I was starving to death I don't think I'd be eating those. Or if I did I'd probably die from suffocation with them stuck in my throat. Not enough water in the world to get those down the hatch. Speaking of water. We've been out taking a walk and it sure is a warm one today. Though I can't say as it was all that warm this morning. Down right chilly. And CindyLou doesn't want to use up our firewood on what she calls good days. Says it's better to be chilly now than frozen to death this winter. But a body's not going to make it to winter, I tell her, if we don't take care of it now. So . . ."

"Help yourself to a drink, JJ. The windmill's pumping and there's a dipper hanging there you can use. I'm on my way to the garden anyway, so come on along. And here comes CindyLou now. Hi there, nice batch of salsify seed you have there."

"Hello Sue. Is that what these are called? I did not remember the name. But I did remember you said they were edible and they sure are pretty. But I hope you will tell me what to do with them."

"It's the roots that are edible, CindyLou. You can buy seed for tame salsify which probably puts out a larger root. But the yellow flowered wild salsify root is edible too. They're not that easy to dig out of the sod of the field though. But you can take the seed from the seed head you have there and plant it in your garden. Then you'll have it right there and it'll be easy to dig and you can see how you like it. Plant a row of it this fall and you can harvest it next fall. Treat it in your garden as you would parsnips or carrots. It's quite a nice looking plant."

"Oh. I see. I *was* wondering how one would eat this large pfoofy thing. I assume those are the seeds in the middle. Not unlike a very large dandelion. I hate to tear it apart just to get the seed."

"If you wait long enough the wind will do it for you. But I know what you mean, it is quite a fantastic looking seed head. And this is seed time. In fact that is what I'm doing now."

"I was wondering what all those paper bags were for. But I did not want to ask."

"Since when don't you want to ask, CindyLou? You sick or something?"

"It's OK, CindyLou. As I've said, I don't mind your questions. I'm collecting a variety of seed today. Some plants mature most of their seed at the same time and I collect the whole plant when it's ready. But some of them like the dill and the onions mature and dry over a longer period. So I go out and clip and collect the seed heads as they are ready. See, I have some in the bags already that I harvested several days ago. You don't have to do it this way, you can just wait until most of the seed head is mature and collect it all at one time. But this way I get more seed. Assuming we have a long enough and a good enough year to do that. Sometimes you just have to get what you can when you can and hope there's enough seed for planting next year. Depends on the plant. We're not in the ideal climate up here for a lot of seed saving. It takes a little planning and care to get a seed crop from most plants."

"But I was reading something the other day that says you shouldn't grow your own seed. That you should leave that to the experts. They insinuated your seed probably wouldn't grow anyway and hinted about harboring diseases and stuff. Now that I think of it I don't recall them saying just exactly why you shouldn't. But they did say you should buy new seed every year from the seed companies. And it needs to be hybrid seed because that's the best. They didn't say exactly why but they just said you should."

"I know, JJ, I've heard the same thing. But quite frankly it's a lot of bull. And marketing I-want-to-get-in-your-pocket propaganda. People have been growing and saving their own seed since the first time man and woman planted a seed in a plot of ground they dug up themselves. Some pretty fantastic breeding and selecting has gone on in those backyard gardens and farms over the centuries. It is still is going on. It's like anything else, use basic common sense and learning and listening and thinking. Then go out and do it. And have fun doing it too. Not to mention a lot of good eating as a result.

I'm not saying you shouldn't buy from a seed company. Especially in the beginning, before you have built up your own stock of homegrown seeds, you will be depending on them for seed. And even after you get involved in seed saving there will be some seed crops that you won't want to bother with, or can't grow the seed in your area. There are many good, reliable, conscientious, usually smaller, seed companies out there. Those that are raising and selling seed for reasons other than just making money. You can tell a lot by reading the catalogues. How honest are their descriptions? How geared toward a particular climate or condition are their varieties? Do they promote and sell open pollinated seed? Do they encourage healthy gardening, are they keeping alive heirloom varieties, do they tell you how to save your own seed?

There are a lot of reasons for growing and saving your own seed. Saving a heritage, preserving and growing diversity, being self-sufficient and independent of outside sources, selecting for special growing conditions and climates, preservation of a variety, genetic health, sharing with others, growing better tasting and better adapted varieties, for food, for fun. Most people probably save seed for various combinations of those reasons. The important thing is that you can do it, any gardener can."

"That sounds very good. And I would like to save my own seed. But I do not know if I agree that any gardener can. I think sometimes I do not know enough to even grow lettuce let alone the seed to plant it too."

"In many cases simply growing food is growing seed, CindyLou. And you are doing very well already doing that even this busy first year with a small plot. You have been eating quite a bit from the land instead of from a box or can lately, haven't you?"

"Yes, you are right, we have been. And we will do more next year. Growing seed and breeding and selecting and crossing just sound so very complicated. JJ was reading to me about all of that, and I do not think I am scientific enough to do it. It do not wish to be a scientist, I just want to be a homesteader."

"Oh, I think you'll find sooner or later that being a homesteader involves many scientific processes and thinking, without you even knowing it. You don't need to worry about it. Seed saving can be as simple or as complicated as you like. You just have to learn a few basic rules. With your love of growing things, and JJ's love of reading and researching, I think you'll find seed saving and back yard breeding will fit into your lives quite well. Let's take a tour of the garden from a seed saving viewpoint, then see what you think."

"Well, I'm not convinced this is something we ought to be doing, but I'm open minded you know. But from what I read, you just can't grow a garden with seed you saved yourself."

"If that's true then you've been eating an awful lot of imaginary food, JJ; and Steve and I have been living on dreams and fiction for a good many years! Not to say that we don't do that too, of course. But we're pretty much living on food from our garden, much of which was grown from our own or another gardener's seed. Look around you. Are you going to believe what you read rather than what is growing right under your nose? Most of this garden is from home grown seed. And the rest is from seed purchased from small growers selling open pollinated seed. Not one plant in here is grown from hybrid seed bought from a large seed conglomerate. If one could only grow food from that kind of seed, then our dinners would be pretty sparse.

"Well now, since you put it that way. Guess you do

have quite a bit growing here. And I guess we've been eating quite a bit of it all summer. Just never thought of exactly where it came from. The seed that is. I suppose maybe one could grow their own seed. But those seed packets in the store are awfully good looking. And there's a sale on them now, too. You can get them pretty cheap."

"And they might grow some food for you, too, JJ. But remember your flowers? Same thing applies to the vegetable seed. It's not that they won't grow something. But it's a matter of what you want to grow. And what you want to eat. And what kind of business you want to support. And what impact you want to have on the land. And . . ."

"OK, OK, I get the point. I'll be careful what I read in the future. Or what I quote anyway."

"It's not what you read, or what you quote, JJ. What matters is what you believe. Whenever someone tells me I can't do it myself, that I have to buy it from them, then I get real suspicious of motives. And with some subjects maybe a little heated up about it too."

"Yeah, maybe."

"I am sure this discussion is important and interesting, to some people. But you were going to tell us how to save seed from the vegetables growing in your garden."

"Of course, CindyLou. I'll just give you an overview. There are some good books on the subject. Johnny's Selected Seeds has an inexpensive booklet called *Growing Garden Seeds* which is great to get you started. And a more comprehensive book, *Seed to Seed* by Suzanne Ashworth, is published by the Seed Saver's Exchange, a book well worth the money. I wish it had been written when I first starting saving my own seed. The Seed Saver's Exchange yearly publications also have many good articles.

Potatoes are easy to propagate. You just save your best tubers and plant them the next year. Throughout the year and during harvest I make note of the healthiest plants. Especially if it's a bad year for blight. That is a great time to select for blight resistance. Save your seed potatoes from the plants least affected by blight but which have a good yield of healthy potatoes. This is where selecting for your own garden conditions and ecosystem becomes quite obvious. If you have trouble with blight you will be especially concerned for blight resistance. If you have scab then potatoes with little scab will be important. If you want a large crop of smaller potatoes then you'll select from plants that grow that way. The same if you prefer very large potatoes, or medium size. You need to look at your seed saving from many different angles, never from just one perspective.

Healthy plants are a good thing. But a healthy potato plant that puts out few potatoes is not much good. Whereas a variety which dies from blight during a bad blight year but puts out a yield which is one of the best of all your varieties is not one to be discarded. It is a lot of give and take, and very subjective. However, the choices are based on quite objective recommendations. Just keep it all in mind, then choose and save what seems best to you.

Now some potatoes will set seed and you can save and plant that seed. Only two out of the fifteen or so varieties I plant usually set seed. The problem with potato seed is that is doesn't come true to the parent. If you plant the seed you will not get a plant and tubers that look exactly like the parent plant. The plants are cross pollinators too. So in my garden with so many varieties of potatoes blooming any seed saved will be crossed and quite variable. Which, to me, is what makes it fun.

You treat potato seed much the same way as tomato, which we'll get to in a bit. You harvest it late fall when the seed balls have turned a light color and feel a bit mushy. It is very fine seed so you need to take care in washing and rinsing it. I've found a tea strainer works well. This small plot of potatoes right here is grown from tubers saved from plants which were grown from seed last year.

Growing the potato from seed is much the same as growing tomatoes. Start them early inside then transplant them out when the weather has settled. Out of twenty four seedlings I planted I saved tubers from the six most promising when the fall harvest came. They were of all sizes and colors but you could pretty much tell what variety crossed with the parent plant since most of my varieties are unique and quite different one from the other. Pink skinned white fleshed, purple skinned purple fleshed, white skinned yellow fleshed, and so on.

This year I planted the tubers which I saved from those six plants and when I dig them in the fall I'll see how they did. When they stabilize to growing out tubers true to form, or very like the parent from the previous year, for three years then I will probably have a stable new variety and can name and maintain it if I want. But I won't bother unless it is truly a good potato. Even if I don't come up with some new variety it is fun.

These **peas** are ready to harvest now. Peas are one of the easiest for seed saving. Grow your peas, pick what you want to eat throughout the year, then let the rest mature and dry on the vine. Harvest them and save the best for seed, then eat the rest as dry peas. Now most of my peas are soup pea varieties anyway so I just harvest my crop in the fall when they are dried and save some of the peas for seed.

Some peas will cross, though they're generally self pollinating. How much crossing you get depends on the weather, your insect population, other flowering plants around that might attract or distract the pollinators, and how large your pea planting is. If you have a farm field right nearby full of field peas you probably won't want to save your own peas for seed that year. Most certainly not if that field was upwind of your garden. But generally you're safe in planting more than one variety if you keep them separated a bit by distance or other tall plants. I don't have a problem with crossing of those marginal self pollinators such as peas and tomatoes and the like. But for insurance I keep them separated as far as I can, given the size of my garden and the number of varieties I plant.

Bean seed saving is likewise easy, same general idea as peas. They are less likely to cross and there is much discussion as to whether what people think of as crosses might just be sports or mutants in a crop. I haven't had any crossing that I know of, and I usually grow more than twenty varieties. Beans are a lot of fun. So many different colors and sizes and shapes! If you can find room for some beans in your garden do plant several varieties. Unfortunately for those with small gardens, to get a decent amount of dry beans you need a bit of space. But at least find a variety of snap bean that will mature in your climate. Then you can eat them as snap beans and leave the rest to mature and dry for seed for next year. And extras can be cooked and eaten as dry beans. Not all dry beans are good as snaps but any snap bean can be eaten as a dry bean.

And the **winter squash**. This is an easy seed saving crop. On one hand it is a simple matter of removing the seeds from a good mature squash before cooking, drying the seed thoroughly, and planting them the next year. However, there is a bit more to it if you want to get a good crop the next year, and many years after that. It starts when you are planning what to plant. There are four main species of squash (maxima, pepo, mixta, moschata) and these four species (in most conditions) do not cross with each other. But within those species the varieties will cross readily. So if you plant zucchini and acorn squash the same year (both pepo species) and then save the seed from either to plant the following year you would get a cross between a zucchini and an acorn squash. Which, while interesting, is probably not what you had in mind.

You can hand pollinate but most gardeners don't want to get that serious about breeding or saving seed. The easier solution is to only plant one variety from each of the four species in a given year. For genetic diversity you want to save seed from a good number of plants so, unless you have a very large garden plot, that will be quite enough squash. If you want to maintain a particular variety this is probably what you will do. Look at the species name as well as the variety name of the squash you plan to plant and simply don't plant more than one variety from that species.

Now if you *want* to experiment with crossing then grow several different varieties of the same species. The results can be fun and you might come up with some great new variety. Most of the results will be edible anyway. This is certainly a good way to breed a variety which is full of genetic diversity. The crop just won't be very predictable. This is what I have here. I'm crossing and selecting to get a good tasting, small fruited, very short season winter squash with a good deal of genetic diversity of the maxima species (such as the Buttercup Squash). I started with four different heirloom maxima varieties of various colors and shapes, all good eating and short season varieties. And each year I save seed from the fruit that best fits what I want. Backyard breeding at its easiest.

Cucumbers are similar to squash, they are all of the Cucurbitaceae family, and different varieties of cucumbers will cross with each other. Generally we eat them when they are immature, so for seed saving you have to let some of the fruits get large and mature before you harvest the seed. (This is also true of the summer squashes.) Growing here is a pickling variety from Denmark, called Spangburg Pickling, which I got from a Seed Saver's Exchange member. Unfortunately, it's not available commercially. But there are a number of good non-hybrid varieties available from the seed companies. Since I don't hand pollinate I only grow the one variety and use it both for pickling and table use, as well as for saving the seed for planting.

Now the **tomatoes**. The king and queen of many gardens and maybe the most discussed crop among gardeners. Up here the tomato topic is generally whether or not you can get ripe tomatoes on the vine from a particular variety or in a particular year. It's a cause for celebration when you do! The tomato is not naturally a northern crop but that shows what can be done with breeding and selection since we grow tomatoes up here. It is another easy seed saving crop.

Let your tomatoes get good and ripe, then when you harvest them for eating or preserving you are harvesting your seed at the same time. Pick the tomatoes for seed from the best overall plants, in terms of health, fruit, earliness, whatever particular properties you are interested in. Cut the tomatoes and scoop or squeeze out the pulp into a glass or other container. Be sure to keep your varieties separate and well labeled if you are saving more than one. Cover the containers with a cloth or loose lid and set them out of the way.

The tomato pulp is going to ferment and this fermenting will help kill off possible disease organisms that are seed born. It should take two to five days or so depending on the temperature. When it is fermenting you will know it. It doesn't smell all that great, the pulp will become covered with a mold, and the seed will separate easily from the surrounding pulp. It's good to stir the pulp a couple of times a day.

When it is ready I clean the seed by filling the glass with water, letting the seed settle to the bottom, then carefully pouring off the water and pulp. You'll need to do this several times. Then pour the clean seed into a strainer, wipe the bottom with a cloth and upturn it onto a saucer. Spread the seed out with your finger and set it aside in a warm, protected area to dry. Again, be sure to label your seeds if you have more than one variety going. When they are dry rub the seeds between your fingers to separate them, then package and label. Junk mail envelopes make good seed packets.

Peppers are also easy seed savers, if you can get them to mature. I have a hard enough time most years just getting a crop. But those rare years when the season is long and warm enough that the green peppers have a chance to turn that beautiful, and tasty, mature red then I save seed. They are cross pollinators so I only plant one variety. Sometimes, when frost is imminent but the peppers are far enough along to show some tint of red, I'll dig up the whole plant and replant it in the bed in the greenhouse. They usually survive and the peppers will continue to mature and give me a chance to get viable seed.

It's hard to fit **small grains** into the typical garden plot. It takes a lot of space to get enough for many meals. And they are not easy to hand thresh. But I almost always plant a few varieties of wheat and hulless barley. Partly to keep some old or rare varieties alive and partly just for the security that if we ever had to depend on ourselves for our entire food crop I'd have some small grains to grow. The small grains aren't prolific cross pollinators but most of them can cross if varieties are grown right next to each other so I separate them throughout the garden. When we had chickens we grew rye. When it was mature we'd use a scythe to cut it down then tie it up in bundles to feed to the chickens during the winter. We'd just untie a bundle and toss it into the coop. They had a great time scratching it around, eating the grain, and turning the straw into bedding.

Broccoli is one of the easiest of the cole crops to set seed because it can be done in one season. Cabbage takes two years and is hard for us to overwinter. You have to get an early maturing broccoli, open pollinated of course, and start it early in the season. It is truly a beautiful plant when in flower. Even if you are only interested in flower gardening and growing plants for aesthetic reasons you should consider the common vegetables.

Many of them are very striking when in bloom. The cole crops cross pollinate readily so you will want to grow only one a year for seed saving. But that isn't much of a problem since well cared for seed stays viable for many years. And most of your cole crops grown for food isn't taken to the blossoming stage. So you grow whatever you want for food and grow a different crop each year for seed.

There are a number of the vegetable crops which, although they are not that hard to grow for seed, take more care and attention. These are the biennials. Plants which grow one year, usually the year we use them for food, then grow their seed crop the second year. For **carrots, onions, beets and swiss chard, parsley, parsnips, and celery** you have to either leave them in the ground for two seasons or dig and replant them the next spring. The good thing about digging them in the fall is you can check the roots carefully and save the best looking for replanting. The bad thing is that some of them are hard to overwinter.

Carrots are not hard. They generally will overwinter even here with a good covering of mulch, and a good snowfall. I pick out the best roots when I dig the fall carrot crop and replant them. If they make it through the winter they will grow a good seed crop the next year with little fuss. And the flower will probably look familiar to you, being a close cousin to the common Queen Anne's Lace, the wild carrot. This presents a real problem for many carrot seed savers, including me. Carrots cross pollinate and are as happy to cross with the wild Queen Anne's Lace as with their tame siblings in the garden. If you have Queen Anne's Lace in your fields you will either have to resort to caging your garden carrot seed crop or plan on buying your carrot seed.

At one time we didn't have much Queen Anne's Lace growing around here and I got by fine by going out and plucking the blossoms off the wild Queen Anne's Lace when they started to bloom, thereby preventing them from crossing with my garden carrot seed crop. I grew a lot of carrot seed this way with just an occasional crossing, which meant an occasional wild and tougher white root in amongst the tame and tender orange ones. Since I wasn't trading any of my carrot seed and was using it only for my own use this wasn't a problem. However, over the years the Queen Anne's Lace population has increased to the point that I've given up growing my own carrot seed. Some day I might build large screen cages so I can once again grow my own seed, but for now that is one of the seeds I buy.

Some of the others such as **swiss chard** and **parsley** aren't quite as easy to overwinter as they often won't survive out in the garden. But I've had good luck with digging the roots in the fall and packing them in a bucket with sawdust. They stay in the root cellar until early spring then are planted out as soon as the soil can be worked. If you do this be sure to save extra roots to allow for those that don't survive. It takes quite a bit of room but you don't have to save seed from every crop every year so you can alternate the more difficult ones.

Onion seed is grown much as the carrot seed is. Pick out your best bulbs when you harvest your crop in the fall and replant them. They are hardy and most usually survive. Our problem is the short growing season. My onion seed often doesn't mature, but when it does I save the seed. Onions are cross pollinators so if you are trying to keep a strain pure you'll want to plant only one variety each year for seed. I've been planting any short season non-hybrid storage onion that I can find and letting them cross. Then I select from those natural crosses striving for an onion that will not only grow a good crop the first year, but will mature a seed crop the second year. This is a long term project.

There are other types of onions, the multipliers and the top set onions. I always grow a crop of potato onions which, while somewhat small, always mature their crop. And as it is the bulbs which you save and replant each year I don't have to worry about maturing and saving seed. They are a good, reliable onion crop for the short season areas, and good keepers too.

Garlic is another non-fuss vegetable and easy to propagate. Hardy too. Just pick out your best bulbs when you harvest them in the fall and replant the cloves. I usually do this the end of September/first of October. While you can plant garlic in the spring and get a crop it won't be as large or as good a one as that from the fall planted cloves.

The last crop is **corn**, and it is a special problem for the seed saver. It's a great homestead crop. You can eat it green or let it mature to dry and grind it for meal or flour. It's a much easier grain to harvest by hand than the cereal grains. You can grow popcorn for popping or parching corn for parching, super-sweet sweet corn or tastes-like-corn field corn. Dried green corn is especially nice cooked with dry beans, or in winter soup or stew. And corn is quite the designer crop. You can get it in the common yellow or white, or in blue or red or purple or pink or brown or black, and even green if you're diligent and patient in your selecting. And in all shades of all of these, even striped and starred and mottled. It is good human food and good animal feed. But it's a bear for the seed saving gardener.

One problem is that it readily cross pollinates. So in the typical garden you can't get your sweet corn crop far enough away from your field corn crop far enough away from your popcorn crop. If you want to save seed you will probably be limited to one variety. In some areas the season is long enough to plant an early and a late corn and be fairly assured of a pure crop if you select from the earliest early and the latest late. And if you have lots of room you can plant your corn in a long block running perpendicular to the prevailing winds, with the different varieties along the row, saving seed from the middle of the blocks.

But the biggest problem is the size of the population

needed to have a good, healthy genetic mix in your seed. At least 200 plants is recommended, saving seed from at least 100 of them. If you want to save seed from more than one variety that is a lot of corn for a small or medium sized garden. Thankfully, for those who can't grow their own seed, there are a lot of good open pollinated corn varieties available from the smaller seed companies for whatever part of the country you live. Even up here.

But if you have a little larger garden you certainly can save your own corn seed. I grow a field corn, mostly of the flour sort, and that serves as both our green and our dry crop. Field corn doesn't have as long a time span for eating green as the sweet corns bred for that purpose, but it is very good eating nonetheless. And in my opinion much preferable in taste. The hybrid sweets and super sweets just don't taste like corn to me. But it is a matter of preference, and what you are used to.

My corn is a cross of many varieties as I'm working on my own selection of a corn just right for our garden and our needs. And I like the diversity in the crop. I don't want a crop of corn in which every plant and ear is just like every other plant and ear. I like to unwrap the husks from an ear of corn and be surprised at the colors inside. And as I'm harvesting by hand it is no problem to me if one stalk is tall and the next short. What is most important is for it to mature a crop in our short season. Second in importance is flavor as a green corn, then easy grinding and good flavor as a dry one. I've found the flour corns to most fit those needs but there is some flint in my cross too as they are the earliest maturing. The diversity of colors in my corn, as with my bean crop, is food for the eye.

You can get information on the number of plants you need to save seed from and crossing and spacing and the like from the book *Seed to Seed*. The Seed Saver's Exchange publications often have useful information. And their annual Yearbook has such a large listing of open pollinated seeds available from seed savers around the world that it will likely awe and overwhelm you when you first see it. But it's a friendly group of people and easy to become involved with. Many of the seed catalogues also include good information on growing your own garden seed.

Growing your own seed is as natural as growing your own food. And I have no doubt that you can do both. You can look through my Seed Saver's Exchange information when you're ready for that. And I'll give you a list of good sources for open pollinated seed, companies which believe in and are willing to go the extra distance to maintain and make available the reliable, important non-hybrid and heirloom seeds. The Seed Saver's Exchange *Garden Seed Inventory* book is a good place to find more companies. But this list will get you started and the seed saving can come as you feel like it."

"Well, at least you didn't come up with anything that I have to build. And I guess you're right that maybe we can grow our own seed. But the whole thing seems a bit much. Maybe I'll just go back and read about it some."

"You can read right after we dig up an end of our new garden plot to plant garlic and onions and carrots. We do not have many Queen Anne's Lace by us so we can grow carrot seed. There is no reason to wait until next year to get started on our seed saving. I think that this is very exciting. We shall start today. And I have been saving orange and lemon seeds and we will plant those and grow our own trees. We shall also plant these salsify seed. And maybe blueberry seed if there are any blueberries left. That way we will not have to go out into the bear's territory to pick blueberries. And I do not see why we can not plant watermelon seed right now. Then it will grow up first thing in the spring and get big and large. I saved some of the seed from that big watermelon you bought the other day. Come along, JJ, let us get going. Thank you Sue, we will see you later."

"Now CindyLou, Sue told you that oranges and lemons don't grow up here, and what's the hurry anyway. I want to get that book and read up some first. We can't just go rushing into something this important."

"She said that orange and lemon *trees* planted here would not survive. She did not say we could not plant lemon and orange seed. You can get that book and find out how. Meantime we had better dig up some of that ground you planted in those oats."

"Dig it up? I just got it smoothed all out and planted! And those oats are coming up too. I'm not going to go digging them up again, CindyLou. Can't you just wait for a few years? You don't have to grow everything this year. You're the one who's wanting us to get at the firewood and complaining 'cause all our clothes are still in boxes, and . . ."

"Say CindyLou? I don't think the blueberry and watermelon and . . . Well, I guess you'll find out. See you two later."

sixteen

The Hunt for Red Potato
and
The Path to the Packed Pantry

~ ~ ~

"We have arrived, Sue, to help you unearth those fresh, white, fluffy potatoes. I am sorry we are a bit late. But I and my shovel are ready. Let us begin the harvest. Oh dear, what, if I may ask, are those? And are you not getting more involved with that dirt than is necessary? It appears to be sticking more to you than to itself."

"I thought you were off changing your clothes, CindyLou? Why'd you take off without waiting for me? Oh, Hi Sue. Kind of looks like you've already gotten a good start on getting into the dirt. What in the world are those things? Anyway, here I was being the ever patient man that I am and I open my eyes to see CindyLou and her lance there head off into the woods.

So then I have to jump up and hurry after her and it's not like I have extra strength to spare you know. You may have to do the digging without me. All this firewood I've . . ."

"You were sleeping, JJ, you were not waiting. And I did change my clothes.

"I was wearing blue you see, Sue, but I decided that since potatoes are so fluffy white that I should change into something more appropriate. And if I always waited for you, JJ, we would still be living back in our apartment in the city. But we are here to harvest the potatoes. Will you be finished with those, well, those . . . items soon, Sue? I am anxious to learn to dig potatoes."

"Glad to have the help, and you look very nice, CindyLou. Although I do wonder whether white might be . . . well, anyway, yes, we are ready to harvest the potatoes. In fact, I've dug a few hills already. The soil is a little damp, ideally it would be drier for digging potatoes. But the forecast is for more rain and it's getting late in the year so I decided to dig them now. The dirt will wash off. Looks like we might have a good day for it too, see, the clouds are on their way out. And that breeze is nice, these potatoes are drying already."

"Those can not be potatoes; surely you are making fun, are you not? They look like . . . well, I do not want to know that they are potatoes."

"Don't want to be crude or anything, Sue, but those dirty colored oblong things do sort of look like, well, they don't look like potatoes."

"I know. These dark blue skinned long potatoes are not the prettiest when they first come out of the ground, especially since they usually have some scab, and especially when the ground's damp.

Some of the blue ones are quite striking when they're washed though, some are even two toned blue and white. But I'm afraid this variety has to content itself with what's on the inside, not how it looks on the outside. But what they have inside quite makes up for their lack of beauty. They are called Purple Cowhorn, and when we're done I'll show you what I mean.

The dark ones are hard to dig because they're so difficult to see in the soil. You have to keep forking through the dirt and feeling around to make sure you have them all. I can see why the commercial growers went to white skinned potatoes. I find all the different colors and shapes make digging potatoes a lot more fun though. They are a particularly fine treasure to seek. See, this next hill are round red ones, they're called Rideau."

"Well, they are certainly quite large and round, not at all like those other ones. And a very pretty red, or I assume they are once the dirt is washed off. I will just take your word for it that those piles of . . . lumps over here are also potatoes. But I had expected them to be tan colored, like the ones you buy in the store. And I did not realize that you had to get quite that far into the dirt in order to harvest the crop. I am very glad we got our solar system up and running, and particularly the washing machine.

"Now will you look at those, JJ! They are the very cutest potatoes I have ever seen! They can not be the same as the large ones you just dug can they? Though they are red colored also. Maybe this digging potatoes will be fun after all. Grab your hoe, JJ, and let us dig in."

"Digging potatoes can be fun I'll agree. And if you hold on a minute I'll give you some pointers so the digging is easier for you. It's best if we get the potatoes out in one piece. And no, this hill is not the same variety as the Rideau. They are called Nosebag and they do tend to be small, at least for me. Types like this are referred to as fingerling potatoes, thin, oblong, crescent shaped."

"Nosebag? What kind of name is that for a potato? Never heard of such a thing. Nosebag. And Purple Cowhorn. Someone should fire their marketing department."

"These are real names, JJ. Named by down to earth gardeners and farmers. The Nosebag potato is said to have arrived in this country in a horse's feed bag. Sure beats names like B-8182-5. Not all of the potatoes I grow are old varieties, a few have made their appearance, and names, in the last twenty years. But those which have been handed down from generation to generation to generation are my favorites, especially if the stories have survived with them. I like to think about someone digging this same kind of potato fifty years ago, or a hundred years, or two hundred years ago."

"Bet they didn't dig them when they were covered with dirt like that. It's going to take some washing to find the spuds under there. And I'm not real fond of washing. Wrinkles the fingers too much. Maybe the next row will come out looking better. Then again, seeing you digging those up and all, maybe it would be better if I got started on the washing, then you wouldn't have to do it all later. Looks like you have an awful lot of rows left to do."

"JJ! We came to help dig potatoes and we will dig potatoes. Sue and Steve have offered us some of the harvest for our winter eating and we will not be poor neighbors about it. So, let us get going. We have to also build our root cellar today, you know, so we will have some place to put all of these potatoes."

"I know, I know. Unruffle. I'll dig. It was just a thought."

"How about if I lend you some other tools? I've found that a fork works better than a shovel. Now this old potato fork probably has some great stories to tell. We got it at an auction and I'm sure it has dug a lot of potatoes in its time. Yet it is still strong and ready to dig more. The tines are longer and closer together than a pitchfork, and there are more of them. Shorter handle too. It works great and is what Steve uses, but it's too big for me, and for you CindyLou. You have to have some larger muscles behind it. You can use that, JJ. And here is a garden fork for you, CindyLou. It doesn't lift the little ones out as well but it works.

"I first push the mulch and vines off the hill I'm going to dig. Then start back here and slowly dig in and under. Some varieties grow their tubers close to the plant and near the top, but others sprawl all over. And if you put your fork in carefully you can usually feel if you hit a potato; then you can dig in farther back and a little deeper. That way you don't stab or cut so many in two. Then pull up with a sort of shaking motion to knock the dirt off and pick out the tubers. Do it again over a little ways, then back a ways. After a while you get to know the varieties and how far out they spread their tubers. But that's pretty much how it's done. You'll get the feel for it once you get into it.

"Now to begin with I want to keep each of the varieties separate, and each plant's tubers separate. That helps me to select next year's seed potatoes from the best hills. So I pile each hill's potatoes together on the mulch of the path as I go down the row. Have to watch your step a bit. Maybe it'll be better if we each work on a different row, probably safer."

"I would say so! Watch where you are swinging that potato fork, JJ! I assure you, it would not be any better an idea to stab me as it would a potato!"

"Just getting the old muscles in shape, CindyLou. As Sue said, it takes some good, strong muscles to handle this big old potato fork. And since Steve couldn't be here to do it, someone has to. Yep, the Farmer JJ and his fork are ready! Let's get to it. I'm just about ready for a snack and drink break so don't be dawdling."

"Uh, CindyLou, as you said, it's better to use the fork to dig potatoes. Not that I don't see your point."

"I was just contemplating, Sue. So, where do we start?"

"Well, this side of that plot has, let's see, Carole, Red Thumb, and Indian Pit potatoes. Why don't you tackle that one. They are yellow, red, and white-with-red-eyes colored. I think you'll have fun digging them. JJ, you can start over there. You'll find a russet, a pink, and a white. I'll finish this row then go to the other side since there are more blue potatoes here, and they are harder to find. The ones in the middle here are called Snowflake and they were introduced in 1873, and grown in this area. Get as many of the tubers as you can. The ones we miss will grow next year and get in the way of the next crop."

"What a beautiful day to dig potatoes. Hello there Carole Potato. My, you a large pretty one. And so many! Look at all these potatoes from this one spot!"

"CindyLou, if you're going to have to talk to each potato you dig up you're going to have to go dig up the neighbor's spuds instead! A body can't concentrate on his work with all that noise. Whew, you sure got that right about needing some muscles to lift this thing. But, hey, look at that! Potatoes! And they even look like potatoes. Like the ones you get in the store I mean. Well, what do you know about that. You grow regular ones too."

"Those are called Russet Rural, JJ. They were once a main commercial potato in the Midwest in the 30's and 40's. They were grown all over the U.P., along with White Rural and Green Mountain. But then different varieties came along and they became almost extinct. Someone must have kept the Russet Rural alive though because a few years ago they showed up in the Seed Savers Exchange catalog. It's good to be growing them again up here. The White Rural, which were even more common, are, as far as I know, extinct. Sad to think of it. But you have to grow potatoes every year to keep a strain alive. Guess no one thought to keep the White Rural alive. The next ones you'll come to are called Linzer Rose, from Linz, Austria. Then a 1937 German introduction called Ackersegen. That one's a good keeper. It can keep your mind busy just thinking of the gardeners and farmers who've grown these same varieties, and the route they took to get to our garden here in the U.P."

"Treasures from all over the world. *Ohhhh, we'll dig the potatoes, find treasures in the dirt...* And look, here are some Gold Finches to sing along."

"Could be they're just coming over to see who the heck's disturbing their peace and quiet! Hey, watch where you're tossing that dirt! I'm getting enough on me without any help from you. And Sue probably doesn't appreciate you spreading her garden soil all over the place."

"I am sorry, Sue."

"No problem, CindyLou. So, do you plan to build a root cellar yet this fall?"

"I would like to have one but we have so very much to do I am afraid it will not get done. And I do not know where we are going to put the potatoes, or the other food I want to get. It is hard to get used to the idea of getting all of your food for a whole year, or for at least the winter, all at once. I know that is what I want to do but I do not know quite how, or where we will put it."

"You'll be surprised at how little space it takes to store enough food for two people for a year, especially if you get bulk food, not packaged and canned. Though our first years we bought, and stored, a lot of cases of canned food in our small cabin. It got us through that transition time between eating out of a store and eating out of the garden. Our main food storage spot was under our bed. All the years we were in the cabin that's where the bulk of the food was kept."

"You must not have eaten much then, there's not that much room under a bed. By the way, when is Steve coming back? Wouldn't want to overstep too much into his part of the potato digging you know. Weren't be very neighborly."

"I don't think you have to worry about that, JJ, Steve won't mind. And I don't know when he'll be back. He had to go to the school to work on a computer problem.

It could be an easy, quick fix, or it could take all day and night. Never can tell. But there can be lots of storage room under a bed, just raise your bed up. Our bed was, and still is, made of wide boards resting on a dresser at one end and nailed to a 2 x 4 nailed to the wall on the other. That's what we started with in the cabin and it worked so well we moved that same arrangement into the house. There's 26" of storage space under there. You can also build a simple shelf in a cool area of your cabin to store your bulk food on, and use a blanket for a door; that will help keep the spot cool. You'll find a place I'm sure."

"You are right, we will come up with a place to store our food. We have talked about an addition on to the north side of the cabin for a pantry. But that will probably not happen this year either. It does seem to take a very long time to put all of these ideas into reality. I had not planned on that."

"You're not alone, CindyLou. I'm afraid if we really knew how long it would all take we'd not get started to begin with. But it doesn't matter, it's the doing that's the fun."

"The other question that I am having, Sue, is what to stock up on. I have been looking through the food coop catalog and there are so very many things listed I hardly know where to begin. And I do not know how much rice or beans or flour or wheat or oranges or tofu we will need to buy to last us through the very long winter. And I do not even know what tofu is. But I did read that homesteaders eat it, so of course we shall."

"Well, you don't have to eat tofu to be a homesteader. It's a good food, but since I can't grow it and have never gotten into making it, we don't. And I think we're still considered homesteaders. You should buy what you think you and JJ will eat, CindyLou. Not that you shouldn't throw in a few surprises, and new and different foodstuffs; but for your main supply you're better off having things you recognize. But you've come a long way in learning to cook and eat real food, CindyLou. You shouldn't have a problem eating well this winter."

"That is true. But I still do not know how much of anything to buy. JJ says we should have a 100 pound bag of rice, but I think that is a very lot. And I want to buy a large bucket of peanut butter but JJ says it will spoil. And . . . say, will you look at this, a dark blue potato grew with these yellow ones. Can that happen?"

"I think you just got a stray from the plant across the plot, there are Yukon Purples growing there. A hundred pound bag of rice is a *lot* of rice for two people, and I speak from experience. A 35# bucket of peanut butter can last just fine if you have a cool spot to store it. We often will buy peanut butter that way because we have more use for the large buckets than the small 5# pails. But you don't want to freeze it. A friend did and he had to chip his peanut butter out and let it thaw whenever he wanted some. But then again, it did keep just fine.

And speaking of peanut butter . . . We started using the pantry/root cellar of the house some months before we moved in, while we were still living in the cabin. The house wasn't quite finished but it was enclosed. It was fall, and we had crops to store. We had gotten our large fall food coop order which included a bucket of peanut butter. This we put in the unfinished pantry. The floor was still just sand but it was cool in there. When we needed peanut butter we'd go over, pull off the sturdy lid, and fill up our jar. One day I opened the lid and noticed something a little funny. The peanut butter had settled with a droop in the middle. I didn't think too much of it, since it always settled back down after having been scooped out. But it seemed like we were going through the peanut butter at a much faster rate than I had planned. It looked as if this 35# bucket wasn't even going to last us through the winter.

The next time I opened the lid to get some peanut butter I noticed the depression was even more pronounced. And the level definitely down. Puzzled I called Steve over and lifted the bucket to show him. There was a certainly a lack of weight to what was supposed to be a heavy bucket of peanut butter. We tipped it on its side and found, to our amazement, a large round hole carefully chewed out of the bottom. And a nice cavity eaten away in the underside of the peanut butter. A very well fed squirrel had found himself quite the treasure. We couldn't believe it, but we had a good laugh. It was an expensive laugh, but funny just the same."

"I think I would be quite upset if a squirrel got into my peanut butter! We shall have to build our root cellar and pantry very carefully if you have trained the squirrels to eat peanut butter. But how did you know to buy that much peanut butter to begin with? I have trouble imagining that much peanut butter."

"I don't think it had anything to do with training the squirrel, CindyLou, he knew what to do without our help. And that was more than ten years ago. We started out with a lot of guessing as to quantity and made a number of mistakes as we learned. At least the mistakes provided humor.

Having a half-mile walk on snowshoes to the road, and car, in the winter started us right off learning to buy in bulk and stock up in the fall. Buying our food that way also allowed us to be able to afford to eat, and eat well. To begin with we bought cases of food from a small grocer. And found out that if you let cans of soup and beans and peaches freeze they are still edible, but they don't look quite as good. And that a case of cocoa, for two people, who don't drink that much hot chocolate and aren't into chocolate deserts, lasts a long, long, long time. After ten years we finally gave away the last cans. The case of powdered milk we purchased wouldn't last that long, and friends' calves wouldn't drink it. But it didn't hurt the compost pile any.

Later we found and joined a food coop which made

buying in bulk easier, and cheaper, with higher quality choices. A lot of things we wanted you just couldn't buy in the grocery store. But the question of how much didn't go away.

Over time I learned by keeping track of how much we used. Each fall I do a food inventory of what is in the pantry. I take last fall's inventory, add what we purchased and canned or dried that year, then subtract the current inventory. That is what we used that year. I made a simple chart with the foodstuff listed down one side and the years across the top. I write in how much we used for that year, and I have an easy, ready reference when it comes time to decide how much to buy or can or dry the next year.

This list was especially helpful when we wanted to cut back on what we bought from outside. We looked at what we were buying and worked at coming up with homegrown alternatives. It's also interesting to look back and see how our habits have changed. Working out, whether one or both of us, made a big difference in how much food, and what food, we ate at home. This is something to keep in mind if you're looking at stocking up on a year's worth of food when more of the family will be at home, and you're used to having them away much of the time.

As I said, our list has changed, and continues to change, as our lifestyle changes. But you can take a look at what our food use was, for two people and a cat, some years ago. We have cut back now on some things such as raisins and peanuts, cheese and powdered milk. And now that we are drying more food some of the items have changed. And there will no doubt be things that you'll want to add, and some you'll want to delete. I know we eat, and are happy eating, more simply than a lot of folks choose to. But this will give you a place to start. When we're done here we can go inside and come up with a list for you.

Since the dirt is sticking to the potatoes so much

this year I think we'll wash them. They can be drying while we work on the food list. In a drier year I just let them dry for a few hours in the sun, both sides, then rub the dirt off as I sort them. Or if it's a particularly bad year for scab I wash them so I'm sure to pick out the most scab free for my seed. It's easier to select the best seed potatoes when they're clean."

"Are you going to keep them all separate like this? It does seem like that would be difficult and require many baskets."

"No, except for new varieties, I mix them all together for storage. Not the seed potatoes of course; they each go in their own marked bag. New varieties I keep apart so I can see how they are for taste and texture when cooked, and how well they store. When we're done we can wash them, either in a bucket or with a hose, select those that we will plant next year (enough for both of us with some extras to give away), then spread them all out on the grass to dry. In very wet falls I've had to spread them out inside on the floor which makes for some crowded living. Especially since those years the dry beans and peas and seed crops are also spread around inside the house to dry. Thankfully this fall is giving us some sunny days for our harvests."

"Well, I don't know as it matters much what you save or store, this old body probably won't survive to enjoy any of it. You sure we don't have enough of these things dug already? A person can't eat just potatoes you know, and these poor overworked muscles surely need something other than just potatoes to bring them back to a reasonable pain-free existence. Muscles don't grow on trees you know. And when you said you'd be digging potatoes you didn't say you were going to dig up acres of them."

"Generations of people have dug acres of potatoes by hand, JJ. And I imagine they were very strong people by the time they were done. But we only have six rows here and the rows are only 25 ft long. But by all means take a break if you need to. Rest your back. It is a lot of work, no doubt about it. But I think I'll keep going. I want to get them all dug, washed, and in the sun to dry in case it clouds up this afternoon. Looks like it could. And it's a lot easier if you can do it all in one day and not have to haul them in and out if the weather turns damp, waiting for enough sun to cure them for storage. It doesn't take much, but some years we don't have much. Why don't you spread out there under the apple tree for a bit. How are you doing, CindyLou, ready for a break?"

"I do not want to stop yet. You go ahead and take a break, JJ. Now look at these potatoes I just dug up! Wipe the dirt off and there they are, with such cute pink eyes!"

"I wouldn't think of taking a break. Can't imagine why anyone who's been digging potatoes for hours on end in the hot sun would want to take a break. And potatoes are not cute. They are potatoes and you eat them, you don't gush over them. And I'm not sure I like potatoes anyway."

"We don't have that many more to dig, JJ, hang in there. The wind is pumping us some cold, fresh-from-the-ground water to look forward to when that last tuber is out of the ground. Though I must say, I don't think it's all that hot. In fact the breeze is turning a bit cooler. Feels like we might be in for a change in weather. And there goes our sun. Yep, look, the windmill is aiming towards the east now. Might not be able to get the potatoes to storage today. But at least we can get them all out of the ground, and maybe washed, so they can be spread out to dry another day."

"There, that's the last bucket. You can put that basket over there, JJ, behind the couch, just leave room to walk between. Too bad about the rain, but we did get them all dug. Thanks a lot to both of you. Do you want to clean up while I get the cook stove going? I think we can use a fire and something hot to drink."

"I think we'd all be better off just standing out in the rain for awhile. If it wasn't so cold. How can it start out so nice and then do this? I think I have half your garden on me. And CindyLou's carrying the other half. You're not going to have any dirt to grow potatoes in next year. Ohhhh, how good a nice hot sauna would feel about now."

"Have to build one first, JJ, they don't grow on trees you know. Just kidding. I admit, it would feel good. Here's a shirt you can change into. It wouldn't have been so bad if it hadn't started drizzling before we were done, makes even our light soil sticky. But falling down into the potato plot probably didn't help you a whole lot either."

"I didn't fall! Those darn spuds tripped me. And every time I went to put my foot down CindyLou started yelling about how I was going to crush some poor little thing till my feet didn't know where to go! So they didn't. And you didn't have to laugh so hard either, I could have broken some bones you know, and then where would we be."

"Here's some warm water; the washcloths and towels are behind the left hand door in that old icebox. And I wouldn't even want to contemplate where we'd be if you'd broken some bones, JJ. Glad you didn't. Want a clean shirt to change into, CindyLou? I really do appreciate all your work. And I'm sorry about your outfit there."

"It is not a problem. They are clothes and they will either wash clean or they will not. I am finding it is not as important as it used to be if they do not. But thank you, I would like to not take all of this dirt with me into your kitchen. And I do want to come in and work on that list of food to order for our winter supply. You said the coop order had to be sent in soon did you not?"

"I'll change my clothes then get my food inventory notebook out, you can look at it while I get the fire going.

And yes, the order needs to be in the mail by Tuesday. If you want I can send yours in with ours."

"Tea, Roma, coffee, hot chocolate? Go ahead and fix what you want, hot water's in the Thermos. We'll have a warm fire going soon. The cook stove is great for days like this when it's not cold enough to have the wood heating stove going but cool enough to want a little warmth. I can get some beans cooking for dinner, too. How about pancakes with applesauce for lunch? It'll be a quick meal, and it looks like JJ is ready."

"Sure am, a man's got to eat after a hard day out in the fields you know. Boy, does that old heat feel good."

"It was only a few hours, JJ, and women have been known to require food also. Pancakes will be fine, Sue, what can we do to help?"

"Not much, thanks. The applesauce is all made and as soon as the stove is hot I'll have the pancakes in front of you. Go ahead and look through the Food Record notebook, the food inventory pages are in the back. The front part is information on canning, a different page for each thing, such as pickles or strawberries or tomato sauce. It's a handy way to keep notes from each year so I know what I've done and what did or didn't work.

"By the way, if you need more 4 or 5 gallon buckets for storing your bulk food in we got quite a few good ones, with lids, from a bakery."

"That is a good idea, we will check it out next time we are in the city. We could use more. There is quite a list of food here, but there seems to be many things missing that is on our grocery list. Do you not buy napkins and bleach and potato chips?"

"With that ton or more of potatoes we just hauled in here why would they want to buy more potato anything?"

"There are a number of things many people would find missing from our list, CindyLou. But over the years we've found we just don't need things that used to be necessities. We have simply come to not use disposables such as paper towels, napkins, sanitary napkins, kleenex - the cloth originals are so much nicer. And we definitely prefer wood or pottery plates and cups. Nor do we find a need for special cleaners or cosmetics, bleaches or laundry aids.

And we've found we get by very well with much less of dishwashing detergent, shampoo, toothpaste, just by deciding to use less. (Actually, we discovered tooth powder some years ago and much prefer that, it's easy to make too. Just mix salt and baking soda, I like a one to two ratio, with a few drops of flavoring oil such as wintergreen or cinnamon for flavor. Mix and shake well and store in a tight jar).

And all of these unnecessities make a big difference, not only to the Earth, but also to our pocketbook. When I look back at our early purchases I'm quite amazed at how much we bought out of habit or because someone said we should. You can view those marketing pitches as a, sometimes, interesting creative endeavor, but you don't have to buy into what they say, or are trying to sell you. It is truly humbling to step back and watch what you buy. But it's also fun to change, and quite satisfying.

You know, I came up with a list of what food we bought/canned/used for a year for an article for *Countryside* some years ago. I bet that would help you out. Let me go find it."

"Yep, here it is. Wow, July of 1991. Didn't seem like that long ago. Back then we weren't eating out of the garden quite as much as we do now, nor drying much food. I think this list will fit your situation better than a current list of our food usage. But you can get that from the notebook too if you want. Interesting difference between 1991 and 1996. It's hard to compare jars of dried food to pints of canned. And remember, this is just what we used of stored food, much of our food is eaten fresh. And we do eat out occasionally."

	1991	1996
Canned Foods		
Snap beans	35 pts	0
Corn	20 pts	0
Tomato (sauce/juice)	90 pts	52 pts
Vegetable soup	30 pts	20 pts
Dry beans*	80 pts	24 pts
Peas	15 pts	0
Tuna	50 cns	24 cns
Applesauce/fruit	40 pts	43 pts
Jam	15 hp	30 hp
Pickles	30 pts	45 pts
Dry Foods		
Powdered Milk	10 #	4 #
Popcorn	20 #	37 #
Dry bean/peas*	30 #	62 #
Whole wheat pasta	40 #	13 #
Wheat flour	90 #	92 #
Cornmeal	3 #	13 #
Oatmeal	70 #	123 #
Brown Rice	10 #	13 #
Barley	2 #	0
Apples	0	8 qts
Green corn	0	3 qts
Vegetables	0	3 qts
Other Food		
Cheese	30 #	5 #
Oil	3 gal	3 gal
Raisins	65 #	5 #
Sunflower seeds	25 #	20 #
Peanuts	30 #	0
Peanut butter	30 #	25 #
Potatoes	1 bshl	2 bshl
Onions	1 bshl	1 bshl
Apples	?	2 bshl
Nutritional yeast	3#	2 #
Maple syrup	10 qts	6 gal
Honey	9 #	1 qt
White sugar (wine)	10 #	20 #
Brown sugar	0	17 #
Non Food		
Laundry soap	7 #	7 #
Shampoo	70 oz	40 oz
Dish soap	33 oz	30 oz
Toilet paper	40 rolls	45 rolls
Cat food	25 #	32 #

"Now before you leave I want to show you the surprise inside these, to some, ugly potatoes I was digging when you got here this morning. Be right back.

"Now, cut this open, and there. Nice, eh?"

"Nice? I am not sure I would say that a potato that is the color of one of JJ's worst looking shirts is nice. Do you eat it? Do you actually grow those on purpose?"

"Purple potatoes?! Who ever heard of bright purple potatoes?! That's just plain weird. And that shirt is not that bad, it's an heirloom. And it's not that bright."

"One can not call an ugly old shirt an heirloom. And the only reason it is not so bright of a purple as this, um, potato, is because it is faded with too many years of washing. It should have worn out long ago. That really is a potato? What does it taste like?"

"Yes it is a potato and it tastes like, well, a potato. When cooked it turns a light lavender. Very pretty in a salad with the yellow fleshed potatoes and the white fleshed and the red fleshed. And all the different skin colors too."

"A potato with red flesh too? Oh boy, I didn't know I was digging psychedelic spuds!"

"Here I'll show you that one, it's this small fingerling called Red Thumb, more pink than red. Depends on the year, some years it's darker than others. It cooks up pink. I think these make cutting up potatoes fun. Every time I cut into one of those purple potatoes it makes me laugh; and the red ones make me smile. And they are good food besides. Why don't you take some home to cook for dinner, I'll pick you out a basket full of variety."

"Well, there can not be much wrong with a potato that makes you smile or laugh. And they are rather, well, interesting looking. Thank you, Sue. Now, let us get on home, JJ, we have to decide what we want to order from the food coop. And what we want to shop for at the next farmer's market. And we have to build our pantry shelf unit. And raise our bed up so we can store our potatoes and onions and oranges underneath, or we have to build a root cellar. And . . ."

"You're pretty big on oranges aren't you, CindyLou? Well, have fun. See you two later, and thanks again for helping dig the potatoes. You can take home your part of the harvest as soon as you're ready."

"Now, CindyLou, this old body's been working mighty hard today, and it's raining, and every smart animal knows that's the time to take a nice little siesta. Naps are important you know, especially when you're a farmer who's just dug an acre of potatoes, by hand, and . . . Purple potatoes. I just never imagined those lumps would turn out to be purple potatoes! Did you? Inside even. They're never going to believe it back home. But when I tell them about digging those acres of potatoes, just me and my fork, why . . . Hey, CindyLou, wait up."

~ ~ ~ *seventeen* ~ ~ ~

Cooking and Baking the ManyTracks Way

~ ~ ~

stomp, stomp, stomp . . . stomp, stomp . . . "grumble, mrumble, snow in November, whoever heard of snow in November . . ."

"Why hello there, JJ, thought I heard someone on the porch. Come on in. Kind of blustery out there today. Where's CindyLou?"

"Oh she's out there admiring this ridiculous white stuff. And she's the one who was so darned afraid of the coming snow. It's not even Thanksgiving yet, how can it be snowing? And it's been snowing for three days already! We haven't got all our firewood in yet. Christmas, yes, there's supposed to be snow. A beautiful white dusting on Christmas Eve, just like on the

Christmas cards. But they never show snow on Thanksgiving! Whoever heard of snow on Thanksgiving. Let alone *before* Thanksgiving."

"Welcome to the north country, JJ. Sometimes the snow starts in October, so you can consider this a good year. Come on in by the fire, don't worry about the snow, it won't hurt the floor. There you are, CindyLou, nice snow, eh? You don't have to take your boots off out there, the stone floor is cold. Come in here on the rug."

"It is beautiful! How wonderful! It is not even Thanksgiving and we are getting snow. Did you see the trees? And the weeds? And the bushes?"

"They've lived here almost twenty years, CindyLou, they're probably so used to this snow they don't even bother seeing it. How can you see it when it covers over your glasses anyway? And it's not wonderful, it's cold, and it's getting all over that firewood I cut up out in the woods last week."

"Oh, do not be such a curmudgeon, JJ. It is beautiful, admit it. Just look out there and tell me that is not beautiful."

"I agree with CindyLou. I've lived in the upper Midwest all my life and I never tire of the exotic beauty of snow. Especially the first good snowfall. No matter when it is. I assure you, JJ, we do still look. And see. Once you accept that it's here and you might as well get on with life you'll feel differently about it. Some good long johns and boots don't hurt either. This is hardly tennis shoe weather."

"Yeah, well I left my boots outside on the porch last night and some darned animal chewed them all up. You'd think he could find something better to eat. They were just nicely broken in too."

"You know that cat that arrived a few weeks ago, Sue? Well she decided to stay and she has picked JJ's new lounging chair as her favorite spot to sleep. He will not shoo her off and he has been ornery ever since. She does seem to know just when he is about to stretch out in his chair and she gets there first."

"I'd throw her off if I wanted to. I just don't want to. And I have not been ornery. Besides, you're going to be pretty ornery this winter when it just keeps snowing and snowing and we run out of firewood because we can't get it in out of the woods."

"You can always sled it in, JJ. Don't look like that, it works very well. Easier on snow than on the ground in a way. And maybe you can patch your boots up. Steve's had to do quite a few repairs on both our boots over the years. Bring them on over and maybe he can help. Glad your cat stayed, she's a nice one. They make quite the addition to the family. You know you could just pick her up and put her on your lap when you want to sit down, JJ. That works. Then you have an excuse to not get up. Because you can't disturb the cat."

"You do not need to give him any such ideas, Sue."

"I suppose not. But don't get too worried about this early November snow, JJ. It often doesn't last and you'll probably get another chance at your firewood. You and half the people in the north are praying for one last chance to get their firewood out of the woods before winter. With that volume and all that whining Mother Nature usually gives in. Though not always. We've had years when winter started in October and stayed until April. That's pretty hard on the old fuel supply. But other years winter didn't set in until into January. Though generally we plan on the end of November or first of December when we can't drive down anymore and we park the car at the top of the hill. That's a nice time. When you know winter is here, you can't do anything about it, and it's time to settle in to the winter routines. I think this is just a temporary winter though."

"Well, I do not care if it is temporary or here to stay. I like it. The wood fire feels good and I am learning to cook on our wood cook stove. Winter is not as bad as everyone told me it would be."

"Just wait until March and there's no sign of

anything green yet. You'll be mighty tired of all this white by then."

"Actually, it's April and May before you can talk about green things. March is still winter. But it's a great time if you decide to enjoy it. You might as well, it's going to be here anyway. How about a fresh baked whole wheat biscuit, just took them out of the oven before you got here. And some strawberry jam? A sure cure for an early snow."

"Thank you, that would be very nice. But your cook stove is not hot. How did you just bake rolls?"

"On the heating stove. Steve made a sheet metal oven that fits on top of the heating stove so I can bake in that whenever a fire is going. It has shelf brackets on the inside and the cookie sheets just fit in. The front opens as a door. See, here it is.

The wooden handles on the sides allow it to be lifted off when it's hot. And this wooden knob on the door twists and turns this piece of sheet metal which is the latch. It works great. And I don't have to burn extra wood to run the cook stove in order to bake in the winter, when the wood heating stove is already going. It takes a little different technique but you get the hang of baking in it soon enough."

"We shall have to make one of those, JJ. Since we do not have a very over abundance of wood this year that would be good for us too. Though I do like to use my cook stove."

"I also love the wood cook stove, there's nothing quite like it. But ours doesn't get used all that much any more. Except for canning time. Nothing can beat the large unencumbered top of a wood cook stove, and all its variations in temperatures. But since we're drying so much of our food now, and eating fresh in season or from stored food so much more, I don't even use it that often for canning. Of course, it's still the best for a hot, quick heat for cooking or for heating the house. We use it more in the fall and spring. In the summer we have the solar oven for cooking, on sunny days. And in the winter we cook on the heating stove, except when it's banked on sunny days when the solar heating panels are working. For quick jobs anytime we have a two burner propane stove we can use. But we only use about one 20# tankful of gas a year."

"We did use our solar oven this summer but then it blew away one day. It was in pretty bad shape when we found it because it had rained. So we have to make a new one next spring. Hopefully one which will not blow away or collapse in the rain. The biggest problem I have is what to cook and how to cook it. But I am learning. I have been using that old copy of *Mrs. Restino's Country Kitchen* that you lent me. And now that the snow has come I am ready to bake. But there is so much information in the cookbook, in all of the cookbooks, that I do not know where to start."

"Would you like a few basic recipes? As you know, I'm not a great chef and I don't spend much time fussing in the kitchen. But I'll be glad to share some of what I've come up with over the years. The biggest thing is to not be afraid to be adaptable. Cooking and baking is truly so much easier than books usually make it seem. Take the basic recipe and substitute when you need to. Experiment with what you like, or what you imagine might work. If it comes out truly inedible then you can just figure you made a special dish for the compost pile. But usually what comes out is edible. It may not be a meal you want to eat again, but that doesn't mean it won't fuel your body just fine. Then next time you try something different."

"I'd just rather we stick to the basics. This old body has managed to survive too many experiments as it is, we don't want to push our luck. How about some plain old food. You know, steak and eggs and hash browns and apple pie and ice cream and . . ."

"Any time you want to make all of that you are more than welcome to, JJ. The cook stove belongs to both of us."

"I know, I know. I made pancakes the other day didn't I? You don't have to make it sound as if I never cook anything."

"The pancakes were a very good meal, JJ. I think I like the program you and Steve have worked out, Sue. He takes breakfast, you take lunch, and you trade off dinner. Though I think I would like to take the entire day off or on. Yes, JJ, I think that would work well for us."

"Well, now, I don't know if I could find time to do all that cooking, I've a lot of work to do you know, and it's not like this old body can just work all the time. Why . . . Oh, unruffle, unruffle. I said I would do more cooking, I'll do more cooking. Now about those recipes? Maybe some ideas wouldn't hurt. Can't say as I want pancakes three times a day, even though I do make some real hunker pancakes, let me tell you. But we don't always have eggs. So I guess I need to come with something else too. How about that bean soup?"

"Your pancakes are great, JJ. But you can make them without eggs. And the soup is easy, you won't have any trouble with it. A nice highlight to our cooking schedule is that Sunday is a day off. No one has to cook on Sundays. In fact, usually we don't eat much, any of us, including the cats. I think it's good to give the body a break, it's healthy. The animals do it all the time. If we feel like eating we eat what's around. Maybe a biscuit and pickles or a handful of popcorn or an apple. The surprising thing is how much more time you have when you aren't preparing, cooking, eating, and cleaning up after, a meal. It's a nice vacation."

"Well now, I don't know about that. We'll have to think about it. Now CindyLou, I don't like that look in your eye. Just 'cause they do it doesn't mean we have to you know. This old body has to have food to get it going and working. You don't feed it doesn't work. And you know how much we have to do."

"Your body will work just fine, JJ, going without food for a day or more. Besides, it's not a bad idea to take Sundays off in other ways too. We don't always do it, but the basic idea is to make Sunday a vacation day. Whatever you are involved in all week, do something different on Sunday. No set schedule, just whatever you feel like. Including just laying around and reading. Or playing music. Or sewing up a bag to put your bike tool kit in. Whatever. Give your mind, your body, your spirit a break. It doesn't have to be Sunday, of course. Pick any day. Just take a day off once in a while."

"But we've come over here on Sundays and you've been working on the same things you're doing all week."

"I said it was a good idea, JJ. I didn't say I was good at it. We have to remember to remind ourselves to take the day off. But anyway, didn't you want some recipes?"

"We would like that. To begin with, how did you make these rolls? I like that they are not fluffy and empty feeling. They feel, and taste, like real food. I would like my bread to be like this."

"Biscuits and rolls bake better than bread in the top of the heating stove oven when you're using all whole grain flour, so that's what I usually make. The recipe is the same though for bread, rolls, or flat bread. And it's a highly adaptable recipe too, all of these are. But to make it easier I'll just give you one version of each of the recipes.

You can buy whole grain flour or you can grind your own. We have a stone bur mill for the wheat (which runs off an electric motor) and a steel bur mill for corn (which runs off arm power). The nice thing about grinding your own is that whole grains keep a lot better and longer than whole grain flour. And it's easier to find wheat than it is good whole wheat flour. Generally you need access to a food buying club or health food store for good flour, but wheat can often be purchased bulk at the local farmers' grain mill. The small grains are usually not treated with chemicals since they are often used for animal feed, but ask to make sure. Unfortunately, most feed mills won't have organic wheat, but it doesn't hurt to ask. It's a growing business. And by checking around you may be able to buy directly from the farmer.

A note about sifting. That is one of those slave-labor and marketing myths. You don't have to sift. We grind our own wheat which turns out quite a bit courser than commercially milled flour. I stopped sifting years ago and have made hundreds of batches of good eating bread and rolls and cookies since.

However, when I'm grinding corn into flour a sifting screen comes in handy. I use one of my seed cleaning screens which fits over a Tupperware tub (which we have). While the wheat can be ground in one step, the corn takes several. It is harder. So I loosen the grinder for the first pass, basically cracking the corn into pieces, but some of it gets ground to flour too. So I sift out the flour and put the rest back in the grinder. Tighten up the burs a bit and grind again. Sift out the flour and pour the courser stuff back into the bin. This is usually a four or five step process depending on how hard I feel like cranking. By sifting I don't have to run that flour through every time.

By the way, barley and oats from a feed store are sure to have the hulls on. Not what you want for grinding into flour. There are hulless oats but they aren't generally available unless you grow them yourself. For oat flour I grind up oatmeal in the steel bur mill. You can get hulled barley through a food coop or health food store.

Now for the recipes:

Whole Grain Bread / Rolls
In a large bowl put 2 cups warm water
Sprinkle 1 tsp baking yeast over the top
Add a bit of sweetening (teaspoon to a tablespoon)
Stir in 1 cup of corn flour or meal
Add whole wheat flour until it is something like stiff pudding
Cover with a damp towel and set aside in a warm place

If you are in a hurry you don't have to let the dough rest and work. I've made very good rolls and flat breads going directly to the next stage. The texture may not be as fine but the eating is just as good.

After a while the dough, also called the sponge at this point, will start working. The yeast will do its thing and the sponge will be alive. How long this takes depends on the temperature, the weather, your flour, your yeast, and probably how you feel. But don't worry about that. How long I wait depends on my schedule, not on what the sponge is doing. I usually start the sponge in the morning and bake in the evening. Or start it at night and bake in the morning.

When you and it are ready stir the dough down and add approximately $\frac{1}{3}$ cup vegetable oil. You can also add:
An egg or two
$\frac{1}{4}$ - $\frac{1}{2}$ cup powdered milk (or whole milk)

Teaspoon or so nutritional yeast

Tablespoon or so ground greens or other vegetable powder

Some leftover squash or other mashed vegetables

Stir it all together then start stirring in more whole wheat or other flour. If you don't have a good portion of the flour be wheat flour then the dough won't stick together or rise very well. That doesn't mean you can't make rolls or flat bread but regular loaf bread might be hard to get baked through properly. If I'm making loaf bread I usually use just whole wheat flour.

When it becomes too stiff to stir with a spoon use your hands. Add the flour gradually as you don't want to get it overly dry by putting in too much flour. This is one of those things you just have to do, you'll get the hang of it. If your bowl is sturdy enough you can knead in right in the bowl, it'll save some mess. If not, use a heavy cotton cloth on your counter or table and turn it all out on that. Keep pushing and folding and working. Add flour until it is no longer sticky. Then work it some more until it smooths out into a nice congealed mass.

If you aren't ready to bake yet you can put this aside in a warm place until you are ready. There is little worry about over rising with whole grain bread. If you want to bake soon, continue on.

For **loaf bread**: Divide the dough into two parts, pat into shape and push into two smallish, oiled, loaf pans. Set these aside to rise in a warm place, probably at least an hour. You can get the cook stove going or dinner cooking while you're waiting.

For **rolls**: Pull off pieces of dough (large egg size or so), roll around in your hand to form a ball, flatten and put on an oiled cookie sheet. Give them a little space between each other, they'll bake faster. Set them aside to rise, maybe a half hour or more. If you don't have time to let them rise put them into a heated oven and bake.

For **flat bread**: This takes more work but there's no need to wait for rising. And it's fast. You need a good hot stove going and one or two heavy skillets. Pull off egg sized pieces of dough and roll out fairly thin, less than an eighth inch, on a floured cloth. Place on a hot dry skillet. Roll out the next one. Turn the first over. Back to roll out another. Turn the second one. Check the first. Etcetera. This is not a job you want to do with distractions around. Basically, you want to lightly brown the bread on both sides. They will stiffen up when they cool. This is a good tasting bread.

The flat bread is something that can be adapted to many different cooking sources, including over an outdoor fire. And you can make them on a not-hot stove, it will just take longer.

Back to the rolls and loaf bread. When they have risen some and feel springy instead of dense pop them into your already mediumly hot oven, whether it be cook stove or solar oven or top of the stove oven. Each will be different for times and techniques, but with them all remember - don't forget you are baking. Charred bread stuff just isn't that much good for anything.

In the top of the stove oven I let the rolls bake until they start to brown on the bottom then I turn them over to finish. The solar oven needs a good sunny day to bake rolls or bread. The cook stove is adaptable to any day and each cook stove will have its own idiosyncrasies in baking. It's important your oven not be too hot or the dough will burn on the outside before it's baked on the inside.

Baking breadstuff takes very little time. You can be fancier and spend more time if you want of course. It's quite adaptable to your desires and schedules. I've made rolls with no rising time at all for a quick bread; and I've had to put my dough in the pantry for two days while I tended to some other adventure. They both made good bread stuff.

And if you need something right away you can make:

Biscuits:
>1 cup water
>1 tsp sweetening
>¼ cup vegetable oil
>whole wheat flour
>1 tsp baking powder
>(optional - egg, milk, half whole wheat/half other

flour)

Mix ingredients, using enough flour to form a firm but slightly sticky dough. Drop balls of dough onto an oiled cookie sheet and flatten with spoon, fingers, or wet fork. Bake in a hot oven until lightly browned.

But what if you don't have, or don't want to use, any leavening? You can make:

Non Leavened Bread:
>2 cups hot water
>1 Tbls sweetening
>2 cups whole wheat flour or other whole grain flour or mixture
>(optional - ¼ cup powdered milk)

Mix ingredients in a heavy bowl, or whatever you have, and set it aside in a warm spot for a half hour or so.

Add and mix in about 1 ½ cups whole wheat flour, little by little, kneading it in as you would regular bread. It won't be as elastic but it will smooth up. Then let that rise all day in a warm spot.

Divide the dough into two and push into two bread pans. Let it rise another hour or so. Then bake in a medium oven for an hour or until done.

Another bread stuff, and a favorite at potlucks, is what I call:

Homestead Chapitas (or Use What You Have Flat Bread)

These are similar to the previous flat bread, they are cooked the same way, but there is no yeast or leavening used, and no setting or rising time. Chapitas don't puff up as the flat bread made from yeasted bread dough does.

Mix 1 cup hot water with whole wheat flour or any flour or combination of flours. Optional extras are ¼ cup oil, cooked grains or cereals, cooked squash, mashed potatoes, mashed other vegetables, etc. If your flour is course the dough will benefit by adding part of the flour to the water, making a thick gruel, and letting it set for a while.

Get the cook stove going. Mix the ingredients, adding flour until you have a soft, workable dough. Put one or two cast iron skillets or pans on the hot part of the stove. Pull off a small egg sized piece of dough, roll it into a ball, flatten, put on a well floured cloth, and roll into a thin round, turning often. How thin you can make them depends on what kind of flour you used. The thinner they are the crisper they will be. If you make them too thick they will be tough to eat and may not cook through. Make them as thin as you can without them breaking apart when you lift them from cloth to pan.

If the dough sticks too much knead in more flour. If the flattened round breaks apart around the edges then the dough is too dry and you need to add water. To do this break the dough apart with your fingers, add a little water and squeeze and mix it all back together, adding flour conservatively as needed until you again get a nice smooth dough.

Put the chapitas, one at a time, on the hot skillets, turning often until they are just starting to brown. Lay them out to cool. They will get crisper as they cool so don't overcook them, take them off while they are still a bit flexible.

It's easy to develop a comfortable rhythm of rolling out, turning, taking off, rolling out . . . Don't get distracted while a chapita is cooking on the stove.

Homestead chapitas are good plain as a snack, as a cracker, as a dipping chip, as a place to spread jam or honey or peanut butter or bean dip, possibly as an edible Frisbee.

Now for something to eat your bread stuff with:

Basic Bean Soup

Cover dried beans and/or peas with a generous amount of water. Cook until just barely done, adding water as necessary. Current year's beans and peas take relatively little time to cook, from two to four hours, depending on how hot your heat source is. As they get older they take longer. Typical grocery store beans must be eons old since they take so long to soften and cook, you had better soak those overnight. Then try to find a source for fresher dry beans and peas for next time.

Add cubed potatoes and other root crops, continue cooking

Add chopped green stuff (fresh or dried) such as parsley, celery, chard

And herbs such as sage, savory, oregano

Sauté briefly onions and garlic, add to the pot

A splash or two of wine, ground dried pepper to taste, a little salt

Simmer all together for a time

Keep warm until ready to eat

And when you don't want beans or peas, or want to fix a faster soup:

Basic Soup

Sauté onions and garlic

Add chopped vegetables of whatever you have and in whatever combination you want. Carrots and potatoes first, greens last, others in between depending on the time it takes to cook them. Or use dried vegetables. Or a combination of fresh and dried.

Add water and chopped green stuff
Simmer for a time
Add herbs, particularly sage
Maybe a splash of wine if it seems to fit
And a dash of salt and pepper
A tablespoon of nutritional yeast can thicken and add flavor to a thin soup
Finish cooking until the texture and flavor is to your liking

Another standby winter meal for us is . . . well, I guess it doesn't have a name. How about:

Bean Goulash
Cook half a cup of dried beans and half a cup of dried green corn in water until done. If you have no dried green (or sweet) corn just do beans. Or add a jar of canned corn when the beans are done.
Drain off extra liquid (save for tomorrow's soup)
Sauté onions and garlic in oil and add to pot
Add a pint jar of tomato sauce

If your tomato sauce is basically chopped tomatoes, maybe with some peppers and carrots and onions, then this dish is good as is. With whole wheat bread or biscuits if you have them. But if the tomato sauce is a thicker concoction then the dish benefits by adding cooked macaroni or cooked cubed potatoes (you can add these to the pot before the beans are done). Or have it plain one day, then add macaroni or potatoes to the leftovers for the next.

In the summer **beans and rice** are a common, and easy, meal. When the solar oven is going that is. I just set a pan of dried beans and water in the oven after lunch, add a pan of rice to cook mid afternoon, and dinner is ready that night. I add whatever flavorings I feel like, but none at all is good too. Plain cooked dried beans are quite tasty. At least homegrown ones are.

Another make ahead meal is:

Potato Salad

In the spring plant a diversity of potatoes, including purple, white, yellow, pink, russet, brown, blue, and red varieties. In the fall harvest your crop and store in a cool spot. Your potato salad will not be boring.

Wash, cube and steam a variety of potatoes (save the water for cooking tomorrow's beans or soup in)

Chop into a bowl:

Garlic

Mild onion, or if strong, chop it very fine and don't use a lot

Pickles

Green stuff of whatever is available or suits your fancy that day

Additions can include apples, asparagus, shredded carrot, about any fresh or cooked vegetable, roasted sunflower seeds, ground dried or fresh chopped herbs

A good **dressing** for the potato (or other) salad can be made with:

      Pickle Juice

      Wine vinegar or apple cider vinegar

      Vegetable oil

      Dried herbs

      Pepper and salt

The first three ingredients can be approximately equal amounts if your pickle juice is good, or it can be just half and half vinegar and oil.

This potato salad is good served warm in the winter as well as room temperature any time. Or cold if you like it that way.

Now for desert or a snack:

Homestead Cookies
$1/3$ cup vegetable oil
$1/2$ cup maple syrup or honey, or brown sugar if you haven't the others, each makes a different cookie
1 $1/3$ cups water (or milk or juice)
2 $1/2$ cups whole grain flour
3 cups rolled oats

Options:
1 - 2 cups dried fruit
$1/2$ cup sunflower seeds or nuts
1 or 2 eggs
Peanut butter
Spices such as cinnamon, or vanilla
Fruit sauce or cooked squash (in place of the water)

Mix all the ingredients together except the rolled oats, adding flour until you have a thick pudding
Let set for a half hour or so for the whole grain flour to soften and, if wheat, develop the gluten
Mix in the rolled oats
Push teaspoonfuls onto an oiled or floured cookie sheet
Flatten with a wet fork
Bake in a medium oven until just done. Don't overcook as these cookies can become quite hard after cooling, depending on the ingredients used. But they are great traveling food; no worry about these guys breaking apart into crumbs.

For another cookie which comes out firm but chewy (unless you cook them to the almost burned stage which is how Steve likes them):

Applesauce Cookies
1 quart apple or other fruit sauce or cooked squash
$1/2$ cup oil (it can also be made without if you haven't any)

²/₃ cup honey or maple syrup (or brown sugar)

3 cups whole grain flour (at least half should be wheat)

Mix together all ingredients

Let set a bit to soften and bind the flour

Drop by teaspoonfuls onto an oiled cookie sheet (this is not as tough a dough as the previous cookie)

Flatten thin with a wet fork

Bake in a medium oven until just lightly browned around the edges

Cool flat on a rack

I hope these ideas will get you going. Good meals don't have to take a lot of time. If you want to spend time and money in the kitchen then that is OK. But if you don't, just decide that simple meals with simple foods are great. That will give you more flexibility and freedom. And you can enjoy the more extravagant fare whenever you feel like making it, or make it a treat to go to a restaurant."

"I think these recipes will do very well for us. I am ready to go bake a homestead meal. Will you look at that! The sun has come out! Now look out there, JJ. Is that not just the most beautiful sight you have ever seen?"

"Sure is, CindyLou. That sun will melt that snow in no time. There's a chance for our wood supply yet."

"JJ! You are not seeing right. But let us get going. We have to make a top of the stove oven. We will need it this winter. And it is time to get that grain grinder out of storage and put to use."

"CindyLou, have a heart! I don't know that much about working metal. And that old grinder is covered with rust and dust. How about we just zip into town for lunch and then we can get to all that. By then maybe the snow will be melted. And we can't wait all day for you to cook up some beans. This old body has got to eat soon or it will famish and not be good for anything."

"You can always pop up a quick batch of popcorn to get you through until lunch is ready, JJ. It's great with nutritional yeast. And I doubt the snow will melt quite that soon, you might as well plan on waiting a few more days. Or just fix your boots and go out and work anyway."

"That is a good idea. I will get the cook stove going and pop the popcorn while you get started on lunch, JJ. I think this is your day for cooking lunch. But I will do the popcorn anyway. The firewood can wait, we have an oven to make. And flour to grind. Come along, JJ, let us get to it. You can not eat lunch until you have cooked it. See you later, Sue, and thank you."

"Now, how do you figure this is my day? You cooked breakfast so that makes it your day. And besides, I have to get to work on finding that grinder if you want to grind flour. And anyway, I'm not sure this is such a great idea, Sue and Steve's ideas aren't always so great you know, and I'm kind of old to be learning how to cook. Unless you want pancakes for lunch. Which you don't so that means you'll . . ."

Solstice

THE SOLSTICE TREE

The solstice tree awakes mid day
To stretch its limbs
Just a bit farther
To feel the touch yet once again
Of both morn and eve
All at once
As all hold their breath
For this magical day
Less than a blink of time
To many

An eon
To others
Where the endings
And the beginnings
Are caught together
Blending
Silent
And all understand
For a brief moment
Before the day breaks away
From itself
And begins anew
It's methodical journey
Into the night's domain
And all celebrate
And cheer
It's growth
As the night
Slowly moves back
Step by silent step
Without animosity
Towards those whose joy
Is in its demise
Knowing it will have
Its day
Once again
When it can wrap itself
Around
And feel the time worn touch
Of both sides
Of the solstice tree
Both morn and eve
All at once.

eighteen

For Love and a Little Money
Life and Livelihood

~ ~ ~

"Come on in, CindyLou. Happy Winter Solstice!"

"Hello, Sue. What a beautiful snowing white day it is out there today. But a bit blustery. One could get lost out there. Oops. Oh, thank you, I do sometimes get a bit lost in my scarf. Stomp your feet outside, JJ, you're getting snow all over."

"Does it ever smell good in here! It'll probably smell even better once my nose thaws. It ever stop snowing here? This keeps on like this and we'll be tunneling our way over instead of snowshoeing. And this old body's not all that much up for digging holes in the snow. Bad enough trying to keep the paths clear. I just get one done and old man winter dumps a bunch more right behind me. It's hopeless. My fingers never get a chance to thaw."

"Hi, JJ. Come on over by the stove and warm up. Some years we only have an occasional storm and some years it seems to snow all winter. This has been one of the snow ones, most snow we've had in a long time. It's great. Except that a lot of snowy days mean no sun days which mean no power. It's been hard on the electric system, as you've probably noticed."

"Sure have! We're back to candles for lights and if we wait much longer for a sunny day to do laundry we're going to have to go buy clean clothes."

"We do not have to buy clean clothes, JJ. There is nothing wrong with washing some things out by hand. It will not hurt you."

"Well, it wrinkles my fingers up something awful. And they freeze faster the more wrinkled they are you know."

"Can't say as I've heard that one before, JJ. But you may not have to worry about it. It's the Winter Solstice. And a great day to celebrate when you live in this climate."

"Well, you did say we are going to celebrate the Winter Solstice tomorrow night and JJ has been telling me all about the planets and the sun and the moon and all of that. But I do not quite see why you are so happy about it. Except that it is very nice to have friends over for a party of course."

"I am looking forward to having friends over, CindyLou, we don't get together that often. But the Winter Solstice is a particularly special time because the days will start getting longer now. Slowly but steadily dawn will come earlier and evening dusk will be later. I don't mind the short days, it's a natural time to slow down and I much appreciate winter for that. But it's time to get a little more sun and light into our days.

"The other reason it's such a time for celebration is because in our area the second half of winter is usually much sunnier than the first half. That means more sun for the solar electric and for the solar heating panels. More power and more banking of the wood stove. Not to mention the positive effect on the psyche. It's a natural time for renewal, for reassessments, for rethinking your life and your living."

"Well now, I'm thinking it's a lot of gray out there and a good time for some long naps. See CindyLou, even Sue's saying you ought to be laying around thinking and dreaming not traipsing off through a blizzard just because you got some fool idea in your head."

"It is not a fool idea and it is not a blizzard. It is simply snowing. Hard. And it is white not gray. And she did not say you should spend all day napping."

"A nap when you feel the need is a great thing, and too often ignored. But I was talking about thinking, JJ, not sleeping. Come on into the kitchen and tell me about this idea you have. You might want to take your sweaters off because I have the cook stove going, I'm making chapitas for tomorrow night. Help yourselves to something to drink and have a seat."

"Can we help in any way, Sue? I do not want to interrupt your work."

"Thanks but it's rather a one person job. Just relax, I can make chapitas and listen at the same time. So what is your idea?"

"We have been talking about what we are going to do for a living, to make money. We saved up enough to get us through this winter but we will have to start earning money next year. Well, I found an article in a magazine that said you could earn very big money by raising this certain kind of pig and selling the babies.

See, here is the article. I do not like the idea of selling baby pigs. But we are homesteaders now and we will be raising goats and chickens and sheep and maybe a cow and maybe a few horses too. Next year that is, once we have our barn built and the pastures fenced. But we could certainly raise some pigs too and then we could make all the money we need and not have to work for someone else."

"Like I said, a darn fool idea. We don't know the first thing about pigs. And we don't have pastures we have fields! With trees and brush and pickers and all sorts of stuff animals don't eat. Now quit glaring at me, CindyLou. We don't even have our cabin done yet. When are we going to come up with a barn?"

"I think JJ has some good points, CindyLou. But the first problem with that scheme is just that. It is a scheme. And this isn't an article it is an advertisement. These things come and go, this cute little breed or that exotic vegetable or some too good to be true work at home get up. But if you look at it carefully you can see what it is. The person who is selling you the breeding stock or the vegetable stock or the manual or the idea is the only one who will make any money. There won't be a market for your overpriced pigs, or it will be a very limited one. You would be simply padding the pockets of a shyster. People who spend their lives scamming and deceiving other people to put a buck in their pockets or another layer of artificial power on their heads have nothing to be proud of.

Sorry, I don't mean to rant, but those things bug me. There are so many creative, fun, and honest ways to live, and to make a living, why pick the lowest you can come up with? But I do understand why something like that would catch your eye. You certainly aren't the first or only one to consider such a scheme, and way too many people buy into it. Now I'm not saying you can't make a living raising pigs, or vegetables, or something else. Many people do, it can certainly be an honest operation. But it is seldom an easy one. And there is a lot to consider before you go buy stock or seed. You certainly aren't going to get rich quick doing it."

"You are probably right. I do see what you are saying. But we do have to make a living and I am not sure what to do. We can work out but what if we can not find any jobs at all next summer? But also I have always worked for someone else. I like the idea of working for myself."

"Yep, working for oneself is the way to go. That way you can take off when ever you want and you don't have to get up early in the morning and you can nap and read all afternoon . . ."

"And you'd be bored stiff before the first month was out, JJ. I hate to burst your bubble but you aren't talking about working for yourself, you're talking about not working at all. There's quite a difference. Besides, you've been "not working" since spring haven't you?"

"Not working? *Not working?* Why these old bones and muscles have been pushed and shoved and worked half to death this year! I can hardly find time to take a twenty minute nap let alone spend all day reading. I've been working harder than I ever did at my old job. And this old body's not getting any younger you know. At night I'm so tired I don't *re-lax,* I *co-llapse.* Not working. Do you know what I did all morning? I . . . What are you laughing at?"

"I'm sorry, JJ. I didn't mean to laugh at you. Believe me, I know what you are saying. I was just reminding you of what you have already found out. Working for yourself can be a hard job, and it does not always mean making money."

"And you did not work *that* hard this morning, JJ. Your muscles are not overstretched. But there has not been very much time off that is true. However, I am not talking about working, we have very much enough of that to do. I am talking about making enough money so we can live here and do the work we do that we do not make money doing."

"Exactly. Livelihood. What will you do for your livelihood?"

"What do you mean exactly? I don't even know what CindyLou just said. But I still say we need to work for ourselves so we can take more time off."

"Working for yourself too often means not being able to take any time off at all, JJ. Or getting so involved in what you are doing that you don't take any time off. It is very easy to do."

"You take time off, don't you? You and Steve go hiking and biking and camping and . . . At least sometimes you . . . You talk about . . . You don't go very often do you?"

"Some years we do better than others, JJ. But we think it's important so we keep working on it. Getting out and involved with being in nature and letting your body be the best it can be is good and healthy and we enjoy it. So we strive to take the time for those activities. It's easy to get involved with whatever you are doing and forget that part of your life.

The nice thing is it doesn't have to take a lot of money or huge blocks of time. An hour for a walk, a few hours for a bike ride, an overnight hike out the back door, an evening with a good book. It doesn't matter if you are working for yourself or for someone else, you can fit healthy exercise and down time in. You just have to decide to."

"That is very good, and I know it is important. Though this year I think my body is getting enough exercise without hiking or biking. And it does feel good. But it will not buy a rototiller or new tires for the car or pay the property taxes. I do not wish to pry, but just what do you and Steve do to make money? I know you do many things and are very busy, but you must make some money somehow."

"That we do, CindyLou. We have always found a way to make enough money. If we wanted more we would make more, but this is all we choose to need right now. Once we realized that an opportunity or job always showed up when we needed it we stopped worrying and just concentrated on living, doing the best with what ever we were doing. That's not to say a job will just land in your lap without you doing anything. It seldom does that. But if you're aware and alert and willing to work at it and follow your instincts then you'll do OK. If you need a job it'll be there. If you're doing what you ought to be doing then the money you need will be there."

"You make it sound so easy. I just don't see any livelihoods lying around waiting to bring us a paycheck."

"I didn't say it was necessarily easy, JJ. Nor did I say it wouldn't be a challenge. Though in a way it is not difficult at all. The job you find right now may very well not be what you want to do the rest of your life. Or for more than a year. Or two weeks. But that doesn't mean you can't do it and make some money while you look for something you'd rather do. As long as you do your best at whatever it is you'll be ahead.

There are certainly jobs I wouldn't take. And there are jobs, and bosses, which I have left for ethical reasons. But there has always been work out there. Both Steve and I have a wide and diverse range of employment behind us. From picking rocks to pulling files. From running a potato truck to running an office. From pantyhose to work boots, from suit to T-shirt. Indoor work, outdoor work, for ourselves, for someone else.

If you look at yourself and your skills honestly, are willing to work, and don't close your mind to possibilities then you'll find a job. But livelihood isn't the same as a job. It can be, but it doesn't have to be. There are lots of jobs you can do to make money, with your head up and your ethics intact. But a livelihood comes from within. It's a melding of your instincts and your interests and your life. It can be one thing your whole life, but often it changes as you change, as your circumstances and your interests and the people in your life change. Livelihood is more a whole than a part. It's a part of you rather than something you do."

"Well now, that's all and good and interesting. But what have you found to make you enough money to pay your bills? What are you doing for a living now?"

"Sometimes I wish I could answer that simply with just one word, or one phrase, one job. But we're all different. Some people do one activity for a living and some do many. It doesn't matter as long as we're doing what suits us. Steve and I are of the many activities group. That's where the name ManyTracks comes from. There are many different tracks in our lives. Some give us more of a monetary living than others, but they are all interesting. To us. Which ones garner the most of our time and attention depends on the season, our current interests, and where we are in our lives. The lineup changes.

Although we've both worked for other people in the past, and there were many good and satisfying jobs, we've chosen to be self-employed for most of the years we've been here.

Probably the most important aspect of our livelihood and employment has been our choice to not need a lot of money. To live, and know we can live, with little outside input. That gives us a great freedom to choose from our hearts, not from our wallets. And it is something anyone can do.

Woodworking has been a long time livelihood for both of us. Steve taught me to carve in wood when we first moved here and we've both been making hand carved wooden spoons since then, as well as Steve's simple flutes, musical instruments, bowls, boxes, sculpture, whatever else strikes either of our fancies. We participated in the art fair circuit for a dozen years and sold our work from galleries. The woodworking was a major portion of our income at one time, sometimes it was a part of it, and lately it's been a minor portion.

Neither of us are sales people. We know the rules

and particulars of selling, but it is not us. We enjoy making, but not selling.

The one thing we have enjoyed in woodworking, other than the creative aspect, is demonstrating. Since a good part of our work has been, and is, done with traditional hand tools we have taken these tools with us to events to share the techniques with folks. It doesn't make much money but it is usually quite satisfying. This, and the nature of our work, led us into the rendezvous circuit - the fur trade era reenactments and living history events. There are a great number of creative people involved as it is by nature a do-it-yourself crowd.

When we started in the art fair show circuit in the early 80's it was a good market. You could make a living doing that, and many artists did. But things changed. More and more art fairs were started and more and more artists were invited to show. The market was flooded. It was a good money maker, for the show promoters, at $100 or $200 per artist and they just kept expanding. And they weren't always too picky about who they allowed in. Production work often outnumbered truly hand made items, and the artist couldn't compete price-wise.

And the market changed. It became harder and harder to pay the bills if you didn't have some less expensive production work to sell. One by one the artists we know have turned to other jobs, other ways to make a living. Sometimes to supplement their work as an artist, sometimes to replace it completely. And these are good artists. It isn't lack of talent or skill that is the problem. It is a lack of a reasonable market.

Not to say there aren't very good artists still making a living with their work. There are. And the art fair circuit is an adventure in itself. I'm glad we did it. But I don't miss it. Having the talent and the skill and the ability is an unfortunately small part of being a successful artist, whether it be in the visual arts, music, theatre, or writing world. That doesn't mean that if that is what is in your heart you shouldn't do it. You just need to realize that it takes a lot more than just talent and skill to make a go of it. If that's what you want to do, then by all means do it. Just don't narrow your vision. Keep your mind wide for possibilities for incorporating your talents and interests into a livelihood.

We still sell at a local coop gallery, from our web site, and at events where we demonstrate. And I added finger-weaving to the woodworking. But it is a minor part of our livelihood now. And this winter a very minor part as both Steve and I took time off to follow other paths. Woodworking is still very much a part of our lives, it is just not a livelihood right now.

Steve's computer work is what is paying the bills right now, and allowing us to spend time on things which don't make much money, yet. And that major source of income has woven its way in and out of our lives for almost twenty-five years. It is a livelihood that was never planned, it just happened.

In 1974 an associate and friend needed a faster and easier way to calculate the myriad numbers involved in planning a client's estate. This was before the change in the estate tax laws and before PC's. He bought an extravagant machine, a minicomputer with 64k of memory and a 150k tape storage capacity. There were no off the shelf programs to buy. If you wanted it to do anything you had to write your own program. So Steve dove into computer programming, teaching himself APL, and wrote a program to do what was needed. It worked. Then we moved north.

About that same time the son of a friend of a friend came north also to open a medical clinic. He was interested in computers for the business end of it and contacted Steve. Thus began a fifteen year relationship with the medical community as computer programmer, computer tech, computer advisor, and, for many years, off and on, as office manager. During that time the use of computers in all businesses, and home life, exploded and expanded and Steve went with it. He built computers, he programmed them, he learned in order to teach an untold variety of software. He ran wire in attics, in basements. He took whatever job was at hand, whatever job needed to be done, and he did it.

He could easily work multiple eighty hour work weeks in the computer business. He came close to it many times. He finally learned to say no, backing off, first here, then there, until he could get the business down to a manageable size. It was interesting, it was challenging, it was often frustrating, but it wasn't what he wanted to do with his entire life.

Right now he is working two or more days a week at the local school as their tech coordinator, or in real terms, as their computer guy. Sometimes it takes more time, seldom less, and they really could use someone full time. But it is a good compromise for both the school and Steve for now. The school is only ten miles away so if they have a problem when he's not there they call and he goes. Most of our out of town trips are in the summer when school isn't in session so it works out well. However, as the job continues to grow it may not be long before he decides to turn it over to someone who can be at the school and available every day. And he would spend more time working at home in the shop.

Steve is often asked how someone can get to where he is, do the work he's doing, the computer work. It's a good job, and it can be taken anywhere. But the answer isn't an easy one. That is, the answer is easy, but the process isn't. He got there by taking on a job, working hard at it, and getting good at it. Then taking on another, usually different, job. Which meant more hard work, more learning. And often low wages. Very low if you take into consideration all the outside hours spent pouring over manuals and learning new programs. It wasn't a matter of luck, it was a matter of keeping an eye open for opportunities and not being afraid to get out there and work at it. But it is still often a surprise to him to find himself with those skills. And a funny juxtaposition to our homestead life. We don't take it for granted, but we do accept it.

The alternative energy business is one we chose to get into, and then out of. We had been living with solar electric for more than ten years and had kept up to date through *Home Power* magazine. There was no one doing that kind of work in the U.P., that we knew of, so we decided it would be a good thing to do. We never got into it deeply, didn't advertise much at all. But what little we did and word of mouth brought us more work than we could fit into our schedule. We put in some systems and it was fun and interesting work. And we met some great people. But it is a business and trade you need to stay involved in, to be good at it. And we weren't willing to put that much time there.

The summer I came up with the name ManyTracks was the last summer we did much alternative energy work. Steve was working out a great deal with his computer work, we were attending art fairs and rendezvous' with our woodwork, I was in the middle of publishing my first novel, running all of our businesses, working in the garden, and we had a number of solar electric jobs going. I was walking back from the mailbox one day and saw my mind and myself following tracks, many tracks, all at the same time, each going in a different direction. ManyTracks became our business name.

We weren't panicking or out of control. We were fitting it all in. But we hadn't put fifty miles on our bikes, and our new backpacks were still clean and shiny and stored in the back room. We knew it was time to step back, take stock, and get our lives back to what we wanted them to be. We enjoyed every track we were on, just not all at the same time. We made some decisions. And have done only a few solar electric systems since. We keep hoping someone will come along with the skills, and the dedication, to start an alternative energy business here in the U.P.

Our lives didn't change immediately with our decision to slow down. We had commitments and promises to keep. It took another year but eventually we got to where we are. Which is still working long hours, but we take more time off now. We enjoy what we do so the labels work and play are blurred. But we read and we bike and we hike. And we rest when we need to. Usually.

Writing and publishing is another part of our lives now, mainly mine. It also is one which was a surprise. Not something I had envisioned ten years ago. Both of us had written articles throughout our years here, mostly by me, mostly about homesteading, and mostly published in *Countryside* magazine. But I hadn't planned on it becoming a business, or a livelihood.

Then one midwinter night I was searching our bookshelves and complaining because I couldn't find anything to read. Which was silly if you took a look at our bookshelves. But it was one of those times I wanted to escape with a novel, and we have few novels, and those we have I've already read, most of them numerous times. I had a particular dream in my head, a short but vivid one, one that kept floating around in my mind. Steve suggested I write it down, write it into a story. So I played with it and ended up doing just that. Before long it turned into a book.

I worked on it off and on for three years or so. At some point I decided to take it beyond just a fun exercise and turn it into a good, marketable novel. One to be distributed for others to read. That's when reality set in, and I discovered the real work of writing. Editing.

At first I assumed I would find a publisher and that would be it. I knew nothing about the writing or publishing world. And it has been a lot of learning since let me tell you. I didn't know where to go for information, had no idea of the vast bulk of information out there, so I just stumbled along on my own. I looked into finding a publisher and soon discovered several things:

One could spend their entire lives and money sending manuscript copies and query letters out to publishers trying to find someone to publish their book. Many people do. But I had neither the patience nor the money to go that route.

No publishers were printing their books on 100% recycled paper, which was a given if I were going to publish anything. It wasn't something which was up for compromise. Either the forests are important or they aren't. To me they are.

I didn't want to turn any part of the creative aspect of the book over to anyone else, including the book production itself.

Between Steve and I we had the skills to do it all. So we decided to do just that. ManyTracks publishing was born. And I knew so little. The whole thing was, and continues to be, quite an adventure and challenge. But it is fun too.

My writing, and the subsequent publishing, is neither organized or of a set pattern. Ideas come when they will, and not when I will. In the middle of writing one book another one demands to be written. An article jumps out unbidden just when I'm buried in book editing. An urge to get a story on paper comes on strong when I have no time for it. Then a play . . . a poem . . . another novel. Creativity seems to beget creativity. The practical inability to do more than one thing at a time puts some order into the chaos.

Being at the same time author, editor, publisher, marketing department, business manager, gopher, and everything else is not something I would necessarily recommend. It's chaotic, it's crazy, it's fun, it's frustrating. When it gets too much I set it all aside and just sit down and write. Someday I may let someone else do some of the jobs; I really would like to spend more time writing. But for right now, I'm the writer and the publishing company. And Steve is coeditor and graphics department.

I did have to set aside the woodworking, for now. And I worry sometimes that I should be doing this, or working on that, or spending time in the shop carving spoons for which there is already a market. But all in all I feel best when I'm following what's inside, and for right now that is writing.

Can you make money being a writer or an author or a small press publisher? For some things I make money, for some things I don't. Being your own publisher is cash out long before there is cash in. And as a writer, unless you write by assignment or contract, your time is spent long before you are paid. And not everything will sell, maybe not much of anything. But if it is what you want to do you do it and do the best you can. Doesn't hurt to have another source of income though.

In the summer my main job is growing our food. And it gets top priority. That doesn't mean it gets all my time, but it does get the attention it needs. I figure I can work out to earn the money to buy food. Or I can work at home and grow it myself. I prefer the latter. I consider it one of my main livelihoods. And Steve always has dozens of projects going - making, repairing, building, inventing, creating - whatever is needed to keep the homestead, and us, going.

Each livelihood has one common thread. If you are going to be self-employed you are your own business manager. Whether part-time or full-time you need to know what you are doing. But it doesn't have to be overwhelming. There is a lot of help out there. You simply have to take the time and energy to learn the rules and keep the business end of your venture(s) out of chaos. It's easier to keep it out than to get it out.

Talking with other self-employed people in the area you want to go into is a great source of help and information. But take time in the library first so you can talk intelligently about the subject. You don't have to know everything before you can ask a question, but you need to know something so you'll know what questions to ask.

The do's and don'ts of running your own business, even a small one, can take volumes. Read what you can, ask questions, search and research for answers. If you go into business for yourself you want to stay in because you want to, or get out because you want to, neither because you have to. But hundreds and thousands of people are doing very well being self-employed, and I'm sure you two can too. If you want to. Don't let go of your dreams, whatever they are.

Employment and livelihood and finding a job is such a personal matter. It comes from the heart and your dreams. What livelihood you end up with will depend on who you are and where you choose to live and what you choose to do. I can give you general guidelines and encouragement. But only you can make the decision."

"And I am afraid you are going to have to make another batch of chapitas for tomorrow night. I am very sorry. I should have made sure JJ ate before we left."

"Well, you said it was only going to be a quick trip over. A body can't exist on air alone you know. And she did set them right in front of me. And you can't expect me to go back out into that blizzard without something in my stomach, I'd freeze to death before I got half way home."

"That's OK, CindyLou, I can make more. I'm glad you both like them. And I don't think you have to worry about freezing to death on the way home, JJ. That light glaring in through the window is called sunshine. It stopped snowing a half hour ago."

"Oh, yeah. I was thinking about my new business and forgot to notice."

"What new business, JJ? I did not know you had in mind a new business? What is it?"

"It's not for public scrutiny yet, CindyLou. It just came to mind while Sue was talking. Now we better get on home. We have some researching to do if we are going to become a part of the self-employed establishment. Come on, CindyLou, get your boots on, we have things to do."

"Now what got you all fired up all of a sudden, JJ? Every time you are in a hurry, rare as it is, we always end up in the library all day. And I do not want to spend tomorrow in the library. It is too beautiful out. And besides, I told you about my pigs. What are you planning? Now do not wrap that scarf rrrnnbb mmbb nnsss! I can dress myself, JJ! Go put your snowshoes on, I will be out in a minute."

"See you tomorrow, JJ. Have fun with whatever you are planning. All set, CindyLou? Looks like you'll have a beautiful walk home. And Happy Winter Solstice!"

"Happy Winter Solstice to you too, Sue. And to Steve as well. And thank you for the ideas. Now I am not so worried about coming up with work or a job or a livelihood. We will find something. I just do not know which one it will be. See you tomorrow. Just look at that sunshine! *Ohhhh, we need no such frou-frou as money or jobbbbbbbs, our dreams will sustain us, we'll dance with the sunbeammmmmmms . . .*"

~ ~ ~ *nineteen* ~ ~ ~

Maple Syrup - Sweet Success

~ ~ ~

"Hi there, CindyLou. Hello, JJ. Happy Spring!"

"Spring? You call this spring!? There's not a spot of ground showing let alone anything green. And it's not easy snowshoeing either let me tell you. If you go out in the morning it's frozen and icy and these old legs just don't skate as good as they used to you know. And then you go out when it's warmed up so you can get some traction and before you know it you're carrying around a ton and a half of mushy snow on top of your snowshoes. And these old legs weren't made to carry weight like that. Why . . ."

"Hello, Sue, it is a beautiful day! And every day I hear more bird songs so they must think also that it is Spring. But even I have to admit it is hard to see it."

"Of course it is Spring. Look around you. Can't you feel it?"

"OK, I'm looking around. What I see are a bunch of

trees with their feet well buried in the snow. And what I feel is wet, cold feet. I don't know why you would want to stand around looking at a bunch of trees in this condition anyway."

"It is a very nice woods you have here, Sue. And I see why you might want to stand here and look at it. But I do not know what it is I am supposed to feel. And if you had oiled your boots last night, JJ, you would not have wet feet."

"This isn't just a woods, JJ, it is a sugar bush. And what you should be feeling, since you now live in the north country, CindyLou, is that true harbinger of Spring - sap season. The trees, the bones, the birds, and the weather, say it is time. The cabin fever cure has arrived. The sap is flowing!"

"I don't see anything special and I don't hear anything special and what are you talking about anyway?"

"You mean it is time to collect maple syrup?"

"Sure is, CindyLou. The nights are cold and the days are warm and there is that feel in the air. We're going to tap trees today."

"Well, why didn't you just say so. Now I can get almost excited about that. That bottle of maple syrup you gave us was good and delicious but it didn't last long. Flooded pancakes here I come! Oh, I can just taste that sweet stuff sliding into my mouth."

"We can collect syrup from the tress around us, JJ. I remember reading about that. You just drill a hole and hang your bottle underneath and wait for it to fill up. Oh, this will be very exciting. To collect our own syrup. Yes, let us go, JJ. If it is time I do not want to miss any of it. We have that large tree right beside the cabin. It should be a very good one to get syrup from. Though I am not sure I want to drill a hole in it. Do you have to drill holes, Sue?"

"Well, yes, but just a minute. There is a bit more involved than that in getting maple syrup."

"Oh, there is always more involved. But I do not want to make a very large amount. And if it is time we can not wait for reading and research. We must go now."

"For once I'm with you, CindyLou. We sure don't want to miss out on collecting any of that sweet syrup."

"You don't have to do a lot of research, but you need to know more about the process. For starters, that big tree by your cabin in a black cherry tree. You're not going to get much maple sap out of that one."

"Of course, of course, I forgot. I know that. But we have maple trees nearby too."

"Yes you do, but many of them are too small to tap. You have to search out your larger maple trees, with trunks at least 10" across. Either red maple or sugar maple will do, we have both here, but sugar maple will give you the higher percentage of sugar. The important thing to know is that syrup doesn't come out of the tree, sap does. You have to boil the sap down to get the syrup."

"Oh yeah, I remember seeing pictures of billowing clouds around what they called a sugar shack. Guess I didn't think about what they were doing. Thought maybe it was some kind of sauna. Hey now, that wouldn't be all bad, taking a nice hot sauna while your sap is turning to syrup. I could get into that."

"I did not want to hear that I have to build another building to get maple syrup. It is not my favorite time to build with the snow still on the ground. And I have a feeling you are going to tell me it will take time to do the sauna syrup part of it and we will not be able to have maple syrup for breakfast. I can tell you are going to say that. You always add more to something that sounds simple."

"Sorry, CindyLou, there is more to it than just drilling a hole in a tree. But believe me, it is worth it. And it's a fun process. It's a great time of year to be outside when there is not that much else going on. But I think we had better start at the beginning. It is simple, really. There aren't that many steps to get to the final syrup. You just don't get to skip any of them. How about if we tap a few trees and I'll explain where we go from there. Steve will be home in a little while and we'll be tapping all the trees then, but it won't hurt to get started."

"I'm with you, the sooner we get to the syrup part the better."

"That would be nice. I would like to know just what we will have to build. It sounds like we should get started very soon if we are to have our syrup for breakfast."

"Actually, CindyLou . . . Well, let's just get to it. You'll see."

"First off we decide what trees we're going to tap. Our sugar bush is small but we only tap about twenty-four trees each year so we divide the woods roughly into two sections. This year we're doing the east half, last year we tapped the west half. Each of us takes half of that, finding our routes from tree to tree, clearing away fallen branches and packing down a path.

Let's go on up to the house and collect our equipment. It's nice that our sugar bush is close to the house, makes it easy to zip out to collect sap even if you're in the middle of something else.

Everything's been washed with hot water and is ready to go. We use gallon plastic vinegar jugs for our buckets. Just cut a hole below the handle for the spile and loop and twist a piece of wire around the handle to hang it by. They're rather unwieldy to store and carry but they are inexpensive and last a long time if you take care of them.

The spiles can be purchased or made; we have about half of each. The handmade ones are 3" pieces of wood dowel or broom handle with a hole drilled through. One end is tapered using a spokeshave so it can be tapped into the half inch hole in the tree.

We need the ½ inch auger drill bit and the hand brace. A hammer, a few jugs, spiles, and nails and off we go, back to the sugar bush. Once the path is packed down and after a cold night we'll be able to traverse the area without snowshoes.

It's good to take along a measuring stick or tool to make sure you don't tap a tree which is too small, it's hard on the tree. Once they're large enough they can handle being tapped just fine, if it's a healthy tree. We don't tap anything less than 10" across. Once they get quite large you can put two taps in but almost all of our trees are in the 10-14" range, one tappers.

Select your tree and decide where to tap. The area immediately around an old hole will be dead wood so you don't want to tap there. We make our way around the tree leaving 4-5" between holes and at a convenient height. This varies depending on how much snow is on the ground when we tap. Some years there is little and we're two to three feet up. But winters like this one, with lots of snow, the taps can end up being four to five feet high when the snow melts. Be sure to look up before you decide where to drill. If there is a large dead branch right above where you want to put your hole move over some. Being below a large healthy branch is supposed to be good, more sap running there.

The hole is drilled about one and a half inches deep up at a slight angle. There, that should do. Hand me a spile and we'll tap it in. You don't have to imbed the thing, just get it in firm enough not to fall out. You have to be able to get them out when you're done. Now take your jug and set it so the spile goes into the hole in the side. Take a nail, put it through the loop in the top of your wire hanger and hammer it in, just far enough to hold. Adjust the nail up or down a bit if you need to. You want to be able to easily slip the jug off the spile but don't want it so loose it will blow off in a wind.

OK, over to the next tree and do the same thing.

Now, this is one of the magical moments of the sap season. Listen . . ."

    . . .   plink   . . .

          . . .   plink   . . .   plink   . . .

... plink ... plink ... plink

"Hey, the trees are dripping! Will you look at that. Do you hear that, CindyLou?"

"I could if you would be quiet, JJ. Yes, I hear it! And, listen, the birds are singing along. Do you suppose they are waiting for the maple syrup too?"

"I think the birds are just happy about Spring. But you will see squirrels licking sap that's running down from a wound in a maple tree. As you can see this isn't syrup, this is watery and it's not very sweet and not at all thick at this point. It takes from 40 to 50 gallons of sap to make one gallon of syrup."

"That is a very large amount of sap! How ever will you get that much? I would think it would drain the trees dry taking that much sap out. Oh, and I was thinking of making a whole gallon of syrup."

"It does take a lot of sap, some years more than others, to make syrup. But the trees can handle it. From our twenty-four trees we make 5 to 6 gallons of syrup, enough for a year's sweetening needs plus gifts. How much you get out of each tree depends on how large the tree is, how healthy it is, and where it is located. Some trees give more than others. Those trees on the south edge of the woods with many branches reaching out into the open usually are the best ones. Generally the larger the crown on the tree the more sap you'll get. That's why trees in sugar bushes are thinned to stand farther apart, so their tops can spread. The weather also has a lot to do with how much sap you will get.

Meantime, we have more work to do. Let's go back up to the house.

We use 4 and 5 gallon buckets to collect the sap so we have to round up and wash four buckets, two for each of us. By now we usually have empty buckets that held wheat or oatmeal or popcorn earlier in the season. You can tell we've used these before because they have gallon gradations marked on the side: 1, 2, 3, and 4 gallon marks. It's nice to have an idea how much sap you've gotten when you collect.

The other thing we have to get is our storage container. We use a polyethylene garbage can for that. It's been holding sunflower seeds for the birds since fall but those are all fed out by now. So it just has to be washed and it becomes the sap storage container.

Now on to the woodshed.

The woodshed doubles as our sugar shack this time of year. By sap season the shed is usually empty inside with walls of firewood stacked around the sides. It makes a handy and convenient sugar shack because it is already here, and it is close to the house. We restack wood as necessary and set up our cooker in the corner.

Before we had our wood shed we made an adequate shack by putting up six temporary cedar posts in a rectangle. We stacked firewood on three sides for walls and stretched a tarp over the whole thing for a roof. It worked fine, though the tarp did get blackened on the end over the cooker.

Now to the heart of syrup making - the cooker. Our first one was made by stacking old bricks on a foundation of cement blocks. A couple of pieces of angle iron across the front and back held up the borrowed sap pan. Some old scrap sheet metal and chimney finished it off. It worked fine except that as the ground thawed the whole thing developed a somewhat scary tilt, and some of the bricks started to crumble. We made about eight gallons of syrup that year though.

But we progressed, as homesteaders do. Steve took an old water heater, cut an almost circle out of one end, leaving a flat bottom, and a rectangle out of one side. Opposite that and at the other end he welded on a large piece of angle iron to stabilize the thing when it's set down. When not in use it is leaned up in the corner of the shed and wood is stacked around it. When sap season comes around we set it back down and make it level using boards, angle iron pieces, pipes, or bricks.

The chimney is made of standard sheet metal pipe and is stored in the storage building when not in use. The cooker/chimney thimble is a piece of scrap metal cut to fit. It is set up to pass just outside the woodshed roof and is wired to it for stability. Where it passes the roof a piece of metal provides a heat barrier.

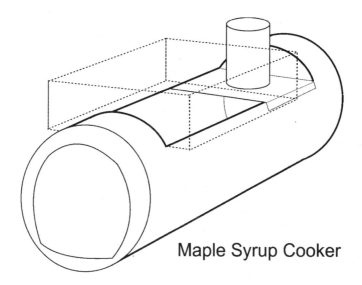

Maple Syrup Cooker

The door of the cooker is a piece of sheet metal propped against the front opening. The draft is controlled by how far out you set the bottom, and a piece of green firewood leaning against one corner is the "latch". A bucket of wood ashes is spread in the bottom of the water heater cooker and it is ready for the first fire.

Now for the pan. For our first try at making maple syrup we used our enamel canner and cooked the sap down outside on a small sheet metal stove. It took quite a while but it worked. The next time we borrowed a nice stainless steel pan from friends who weren't cooking that year. Then we made our own.

We have to go inside and get the pan down off the wall where it hangs as a typical homestead decoration most of the year. Watch out there I don't bump you with it. Not that it is fragile or anything but it's covered with oil. Steve made the pan from a piece of 36" x 48" 24 gauge sheet steel; a thickness he could bend and braze without special tools. The top edge is folded out and down 2" for strength, and four heavy steel handles are bolted on. The final size is 18" x 28" x 8" deep. It is just wider than the rectangular hole cut in the side of the water heater. And is a size that is manageable for pouring out the syrup once the sap has boiled down.

At the end of each season the pan is cleaned then oiled inside and out. After almost ten years the pan is well used and far from shiny but it is still doing its job well. I never completely scrub the outside, just get most of the black off so it doesn't get all over (crumpled newspapers work well for this).

Now that the season has begun, the twice a day (usually) collecting routine becomes a part of our lives. Off-the-homestead trips are planned around how the sap is flowing. And it feels great to be working outside again.

We collect sap until the storage container is full then we start our fire. A reminder I want to give you, from experience I might add, is to be sure your pan has sap in it *before* you start your fire. We fill it up then boil the sap down to about ½" to ¾" three times for each batch of syrup. When we're boiling we try to time it so there is an inch or so left in the pan at night and the fire is going down. The embers will gently reduce the sap over night without fear of burning.

We've found that the best wood for the sap cooker,

and the most convenient for us, is dead poles (tree tops and branches). When we're collecting firewood in the fall we also collect poles to pile up beside our woodshed. Some of the poles are cut to size for the cook stove but the rest are saved for sap season. We cut them to length by hand as we need them. Since the sawbuck is right there it's an easy and enjoyable chore. It seems to fit the whole process.

When you're boiling you'll find a lot of foam and stuff coming to the surface that you'll want to keep skimmed off. Not only for the quality of your syrup but to keep it from insulating the sap and preventing a good evaporative boil. We've found the easiest skimmer to be a length of thin light wood, such as $1/8$" x $3/4$" pine, the width of your pan, with another piece nailed on perpendicular for a handle. You just float it across your sap and scrape the scum off into a pail. It's easier than using a spoon.

When the third fill-up has boiled down, the sap (almost syrup at this point) is poured off through a clean flannel blanket into a stainless steel or enamel kettle. I finish it off inside on the cook stove or wood stove. If the wood stove is going I set it on that and let it gradually cook down to be syrup. Or I'll get the cook stove going and do it faster there. You can either go by feel as to when the syrup is cooked down enough (how it drips off of or sheets across a large metal spoon) or boil it until it reaches seven degrees above the temperature it takes for water to boil.

When the sap is a good syrup I bring it to a boil and strain it again through a piece of wool blanket and a cotton cloth (such as T-shirt or flannel blanket). Then it gets bottled in clean jars, bottles, or jugs. If while in storage a mold occurs on the syrup (as I've found an occasional jar or batch to do) just pour it into a pan, skim, boil, skim some more, and rebottle in clean bottles.

So there you have it - sap to syrup. And it's the finest sweetening you can get (in my humble opinion).

We use it for everything. From cereal to jam, pancakes to pickles. I've even tried it in wine making but I wasn't thrilled with the results. If I didn't have sugar it would do though. And I don't think I've found a more appreciated gift than a bottle of fresh from the woods maple syrup."

"Well, we have work to do, JJ, let us get going. The afternoon is waning and we have to build a cooker and get jugs and make spiles and make a pan and find a drill and build a woodshed and . . ."

"Now hold on there, CindyLou. I'm as fond of maple syrup as the next guy and I sure was looking forward to having maple syrup smothered pancakes tomorrow morning, but no one warned me it was going to be so much work. This old body won't be around to appreciate the sweet stuff if you work it to death first. It's not getting any younger you know and a body's only got so many projects in it. Now that I think of it, that store bought syrup stuff isn't all that bad, and . . . Oh all right, unruffle, unruffle. I suppose we could come up with some way to make a little syrup this year . . . we do have that large canner we can boil it in . . . and maybe we could make a cooker out of that old steel drum sitting out back . . . and we have a broom we can cut up for spiles, who has time to sweep anyway and . . ."

~ ~ ~ *twenty* ~ ~ ~

Great Expectations - Small Greenhouse

~ ~ ~

"What do you have there, CindyLou? Hi, JJ. Rather a windy one today. Oops, watch your, uh, plant there. Here, I'll get the door."

"Thank you. There *is* a bit of a breeze. And I am afraid my tomato plant did not appreciate the walk in the wind very much."

"CindyLou's got these leggy plants all over our window sills and she's got the idea they are going to grow tomatoes and peppers and melons and I don't know what all this summer. As far as I can see all they're doing is blocking the windows."

"You are the one who wanted melons in the garden so do not complain because they are taking up space in the house. If you want to see what is outside then you can simply go outside. You do not have to blame the plants."

"It looks like your tomato plant is a bit under the weather, CindyLou."

"Probably because she was yodeling to it all the way over."

"I was *singing*. Plants like to be sung to. It is a good thing you are not growing our plants, JJ. But I think maybe my singing has not been enough. Or I am singing the wrong song, I do not know."

"I'm sure the plants like whatever you sing to them, CindyLou. It's what's behind the song that makes them grow. But where do you have your plants growing? What window was this one in?"

"Well, we do not have enough window sill space and I am using them all. This one is in the window looking out into the woods behind the cabin. But it does not seem to like it there. Nor do the others."

"It's because it's a north facing window, CindyLou, and your plants need more sunlight than they can get there. You need to use your south facing windows for growing garden seedlings; there just isn't enough light coming through the others."

"But there is not enough room. We do not have enough window sills which face south. And I have many plants to grow. And they are getting bigger all the time."

"It is a problem, CindyLou, I agree. Even with the greenhouse our south window sill space is crowded in the early spring. We added an extra shelf inside to help out."

"I have been thinking of your greenhouse. I did not know before why you would put such a thing on the front of your house and block out so much of your windows. But now I understand."

"Now CindyLou, I don't like the look in your eye. Every time you get together with Sue I end up having to go build something. And it starts with a look like that. Doing that maple syrup was enough for one spring don't you think without coming up with something else right away? A body's got to rest some time you know."

"Do you want to grow vegetables this summer or do you not? If we want to grow our own food we need to grow our own plants. And that is what we will do. Now, about your greenhouse, Sue. How do we build one? We do not have much room in front of our cabin. But we could do something on one end maybe."

"Well, CindyLou, the greenhouse has been very handy. But before we had that I grew many plants using the windowsill and a large cold frame out in the garden. I started the tender plants inside then put them out in the cold frame when it got warmer. That way they got more light, had more room, and got used to being outside. The hardier and later plants I started directly in the cold frame. It was a simple structure built of used cement blocks, old windows and boards. It was five windows long so it gave me quite a bit of space. It worked well and I used that combination for many years. Once I had the greenhouse though I found I didn't use the large cold frame so we dismantled it. But something similar might do for you until you have time to plan and build a greenhouse."

"I was not in the mind to spend a lot of time planning. Surely a greenhouse can not be that complicated."

"Greenhouses come in all shapes, sizes, and complexities. It can be as simple as a board or pipe framework with plastic draped over or a complex glass and stone structure with a pool inside. It can be attached or freestanding, little or large, above ground or half below. It all depends on what you want and how much time and money you want to spend on it. We have some books we can lend you that will give you some ideas. The structures are as varied as the builders, as is the use made of them.

"Time spent planning and thinking will be worth while, CindyLou. You want to end up with something that is not only usable but enjoyable as well."

"Well, the books and planning will make JJ happy.

I just want a place to grow my plants. If we build a greenhouse I will use it, and I will enjoy it too. But I suppose I do need to learn about it first."

"We get a lot of use out of ours and it is not very large, only 8 ft x 13 ft. It is an attached modified pit design. It was built with a lot of used and leftover material so it didn't cost much. We designed it around the windows and glass that we had. I especially like looking out of the shop and into the greenhouse in the winter and seeing green things. And since it covers half the area of the shop windows it gives us a glare free spot to work when the winter sun is shining straight in. And being attached to the house helps keep the greenhouse warmer since we don't heat it. The entry to the greenhouse is off our house entryway so we don't have to go outside to get into it, which also keeps the cold air out in the winter.

Since the front of our house is low in the ground we made the greenhouse low too, but not quite as far down as a regular pit greenhouse. The floor is a step down from the entryway level. Only the front half of the roof of the greenhouse is windows, the back half is sod, which blends into the sod roof of the house. Since we don't need the sun and heat in the summer, when the sun is high overhead, the solid portion of the roof provides welcome shade, and in the winter insulation. In the spring and fall the sun path is such that the sun starts shining through the glass portion and into the greenhouse when we need it, and all winter, too. We angled the roof to take best advantage of the spring and fall suns, the most important time in our greenhouse.

Along the front windows is an 8 foot long by 28 inch wide slatted wood bench which wraps around the west end. There is plywood directly underneath, slightly tipped, to direct water and dirt towards the aisle. Underneath is storage for pots and tools.

Along the other, house, side is a 9 foot by 28 inch waist high bed using Oehler's (*The $50 and Up*

Underground House Book) PSP method (post/shoring/ polyethylene). Cedar posts were set in the ground, boards (shoring) were laid up against the posts, and polyethylene (10 mil is nice or two layers of the more common 4 mil) is carefully laid against the wood. You can lay some roofing felt on the wood first if your boards are rough.

This bin was then filled with regular dirt (be sure to remove sticks and sharp stones to prevent them tearing the plastic). We tacked scrap boards around the inside top to protect the plastic from trowels and such, then added a board across the posts for a handy ledge.

This bed turns out to be especially useful in the fall. I bring in full sized plants from the garden, transplant them into the bed, and have an instant inside garden. It also holds heat on a sunny day to help temper a cold night.

The east end of the greenhouse, where it opens into the entryway, is where the garbage cans fulls of compost (potting soil) and flats are stored. The floor is screened 1"-2" stone which is good for drainage, but hard on bare feet.

The greenhouse has its own life and patterns.

Winter is fairly quiet as I visit only occasionally to harvest a few green leaves for dinner, and to say Hi. Late winter finds more activity as I begin to rummage for seed pots and sift compost for seed starting soil. Seeds are sorted, the garden plan consulted, the greenhouse notebook dusted off. Tin cans with holes punched in the bottoms make good seed starters. I pull out an old cookie sheet, line up my tin cans, attach new masking tape and pencil labels where needed, drop in a layer of small stone, fill with sifted compost, and carefully count out the seeds. A little more compost over the top, a gentle watering with warm water, let them drain, and cover all with a newspaper. These earliest plantings come into the house by the wood stove to start. Then they move to the kitchen window, replacing the cat and a few house plants. Later they move into the greenhouse.

Spring is the most crowded and busiest time of the greenhouse year. The bed still has winter plants growing but they are chopped down as they (generally) start going to seed, and more room is needed for flats. More seeds are started, seedlings are transplanted into flats, some small plants are transplanted again into large flats and pots. There is not an inch to spare in the greenhouse, and the garbage cans full of compost are empty. And I search for boards to make just a few more flats.

Flats are easily made of scrap wood of whatever type is available (not painted or stained with toxic coatings). My basic flat is 12" x 16" x 3" made of ³/₄" pine or poplar with holes drilled in the bottom. The pine is lasting better than the poplar, but we had poplar so that was what I used for half of my flats. Drywall screws work nice for fasteners if your wood is apt to warp.

Most plants go from the seed cans into a regular flat then into the garden. But I also have some large flats for the second transplanting of tomatoes and peppers. They are 14" x 16" x 7" and 18" x 15" x 5" and I wouldn't want them any larger. Filled with dirt these critters are *heavy*. But they take up less space than pots and are quite handy. I've also used them for winter plantings of greens.

Spring is also the time for the most intense management in the greenhouse. We have openable windows along the front and sides, as well as upper and lower vents opening into the house. These, along with an insulated curtain on the front windows, help me to keep the temperature from the too hot or too cold. Guessing what the weather will do while you're gone is a game of frustrating futility. I've learned to opt for assuming sun as overheating does more damage than coolness this time of year. If we're to be gone for long I lower the front curtain and open the vents into the house.

An early addition that helped a lot for the health of the plants was to install fans. You don't need a heavy breeze, just a gentle moving of air. For ours Steve hooked up two large muffin fans (from dead computer power supplies) in opposite corners of the greenhouse. He installed an old thermostat (upside down) so we can have the fans come on and off automatically as the temperature goes up and down. There is also a manual switch. In the spring the fans are on almost every day.

As spring turns to early summer the flats spend more time outdoors, getting acclimated for the coming move into the garden. A lot of trips in and out and out and in. But they do much better in transplanting when they've already gotten used to being outside. I start them in a shady, sheltered spot to begin with then gradually move them into the open.

Then, suddenly it seems, the greenhouse is empty, and the garden is full. Summer has arrived. Though there have been cold, rainy summers when some plants spent all season in the greenhouse. But generally summer is a time of rest in the greenhouse, when the pots and flats are stacked aside and the area is left to the resident spiders.

Early fall once more finds activity as spiderwebs are swept away and the resting bed is dug up. As frosts become more frequent in the garden I start bringing in plants. Although I've experimented with many things I usually transplant in only swiss chard, parsley, and celery. These have been the most successful for me for overwintering as we don't heat the greenhouse in the winter and some years it gets quite cold. I prune off dead and outside leaves and bring in the healthiest, smallish plants.

Some years I've dug in full size pepper plants too, full of green fruit. Most of them survive and go on to ripen their peppers giving us fresh peppers into early winter. These generally harbor aphids, though, which tend to multiply unchecked in the greenhouse so the plants are pulled and put in the compost pile as soon as the peppers are gone (the best solution I've found for an infestation of aphids). Some full grown marigold plants are a fun, if temporary, addition also. They do a good job of brightening the greenhouse until it gets too cold.

Fall also finds our greenhouse full of drying corn and bean pods, spread out on the front bench or in bags or baskets, wherever room can be found. I've also thought it might work for drying apples if you didn't have a dryer.

To have winter salads the plants, no matter what they are, need to be close to full grown when winter sets in as little growth is made once the days shorten. When I grew lettuce and greens for winter I planted the seed in mid September and transplanted up into the large flats or into the bed as they grew. I found lettuce in general to be disappointing except for varieties bred, or selected, particularly for gardening under glass, such as Diamante, Salina, Marvel of Four Seasons. I had better luck with other greens such as Kyona, Pac Choy and leaf Chinese cabbage.

A fun addition to the late winter inside garden has been to plant some spring flowering bulbs in the back of the bed in the fall. It's quite a treat to have bright daffodils popping out here and there while the world outside is still gray and white. And I usually let a few Johnny Jump Up's stick around and flower, too.

An additional management of the greenhouse in the winter is the use of insulated panels and curtains for the windows. This has done a lot towards keeping the heat inside at night and on cold cloudy days. The curtain on the front windows is also used for shading in the warmer months.

The panels for the side windows are of simple construction, made of a 2" x 2" wood frame with rigid foam inside. Polyethylene film is stapled over both sides (to keep the outgassing from the foam contained), then large cardboard pieces are stapled on. They are painted with white oil paint to protect the cardboard from moisture. The light color is important for light within the greenhouse and to reflect heat out to avoid overheating the panels when the sun hits them.

The curtain is a simple quilt sewn up from white and natural colored scraps of cotton material, with quilt batting between, and tied every four inches. The quilt is attached above the windows, and wooden strips are stapled and screwed on the bottom. It is raised and lowered with light rope and pulleys in a manner as you would a rolled reed curtain. The insulated panels are held up with various latches, toggles and bungie arrangements to fit the different windows.

Another good addition to the greenhouse was glass and plastic jugs and bottles filled with black liquid. Placed mainly where the sun can hit them along the front and side benches and on shelves filling the area between the posts of the front bed (which was painted black) they do a great job of tempering the extremes of both hot and cold. The liquid was made with black RIT dye, $1/16$ tsp dye per quart; $1/4$ tsp per gallon. One package colored 23 gallons of water. I also added some old salt to help lower the freezing point of the water. In the cold of winter I move the jugs away from the windows; and in an especially cold period I bring them inside.

For seed savers the greenhouse can be especially helpful for starting difficult plants. It gives you room to start and grow plants needing a particularly long season to set seed. You can also overwinter tender biennials such as swiss chard, celery, and parsley, setting the full grown plants in the bed (or large container) for the winter then transplanting back out in the garden in the spring to grow seed the following summer. It is a good place to overwinter tender herbs also.

The greenhouse is not an essential tool but it sure does help the short season grower. I got by well enough for many years with just the large cold frame, but I do appreciate my greenhouse a lot. It's not fancy, but it's friendly."

"That is a very much to do but this tomato plant is worth it. We will have to build us a greenhouse, JJ. Maybe we can borrow your books, Sue, and start looking at what we want."

"Be glad to lend them to you, CindyLou. One of my favorites is *Winter Flowers in Greenhouse and Sun-heated Pit* by Kathryn Taylor and Edith Gregg. Even though I only grow an occasional Johnny-Jump-Up or Marigold in my greenhouse I got more ideas and plain good information out of this book. And there are many books in the library to look through."

"I'm glad to hear you want to study this some first, CindyLou. We have enough to do without building something right now."

"We can not, of course, build our greenhouse yet, JJ, there is still snow on the ground. But we can build flats. That will be just what we need to fit more plants in our windowsill, which we will have to do since we have to move all of them to the south windows. And as soon as we can see the ground we can build our own large cold frame. I saw an old fallen down building made of cement blocks lying all tumbled around not that far away. Maybe we could get some of those blocks. And we can build a shelf across our south windows just to the size of our flats that we are going to build.

"Come on, JJ, we have things to do. Our tomato and pepper and melon and orange plants need to get in the sun. Yes, look at our tomato plant perking up already. I am sure he is going to be just fine. And he will grow into the very lushest tomato plant you have ever seen, just full of beautiful ripe tomatoes. It is going to be a great summer for our plants, JJ. See you later, Sue.

"Ohhhh, the tomatoes are coming, make way for the viiiiiines, the melons and orange plants need a place to grow toooooooo . . ."

"For heaven's sake, CindyLou, that doesn't even begin to rhyme, even I can tell that, and if whoever wrote that song ever heard what you were doing to it, why they'd . . ."

"Orange plants? I thought we already talked about that, CindyLou? Oh, well, see you later. Orange plants. Well, you never know . . .

"Hmm hm hmm hm hm hmmm, their chickens will crow, their bulls will have calves, their orange trees will blossom, midst winter's cold blast . . . hmmm hm hmm hm hm hmmm. . ."

~ ~ ~

Dream Your Life, Live Your Dreams, and be Prepared for Overlap

~ ~ ~

Try this. Create in your mind the world as you think it should be. Go ahead and dream of your own utopia. People it with friends. Endow it with a healthy Earth. Detail it with life and liberty and happiness for all. Include all the creatures you want no matter how endangered. All the forests, all the clean seas, all the clear skies. Whatever you want to choose.

What if you choose dirty, crowded, crime filled cities and power and greed? Well, you have to choose what is most important to you; what you want. But I'm not too concerned about that happening. If you picked up this book, and have read this far, you aren't likely to be a person who chooses destruction. Besides, I'm working on my own future dream world. And in that world people predominately choose creative dreams and actions, not destructive ones. It doesn't hurt to bury your head in the sand once in a while; it clears the mind.

When I was in the midst of work on my novel, *The Last Lamp,* I became quite fond of the people in the book. They became friends whom I missed when I came out of their world and into my own. I also developed a real homesickness for the world where they were living. It is not a perfect world, they all have their problems to work through, challenges to meet, just as you and I do. But they have gone beyond some of the problems that are so frustrating to many of us today. They take the honest living with nature and being a part of All That Is/God/The Great Spirit as a matter of course. They can't understand how we operate in such a destructive manner in our world. When I left, I missed them. I missed their world.

Then one morning a thought skidded into my head - *That world will never be if you don't start living today as if that world, and that society, existed today.*

If I don't . . . It didn't say think about it, rant about it, write about it, it said *live* it. That world will never be . . . This wasn't just some fantasy story, some fiction that didn't matter, to be forgotten the minute I put the book down. These were my friends. These were people I had come to love. This was personal. The way I lived today, the decisions I made, the choices I picked, the paths I took, would affect whether or not my friends ever got to be. Whether that world ever came to exist. Or, if I live today as if the world were a better place, then it will be.

My life changed a little with that thought. My decisions took on a new light. The seemingly insignificant choices didn't seem so mundane. The frustrations less important. This I could do. First dream the world I wanted, then live it. Now. Today. Mmmm. So simple. So difficult.

If you sometimes wonder if you are being too fanatic about caring for the Earth, about being a part of Nature, about living a healthier life, think not only about what you are doing for yourself and those around you, but also for those coming after. What kind of world are you making for them? Or for yourself. How important is it?

The nice thing about this is that it is fun. It doesn't need to be a downer poor me martyr thing. Anyone can afford to do it, anytime. No hype, no bragging, no ranting. Just plain living your life the best you can imagine it being. The way you would like it to be.

It's important. So go ahead and spend time on your future dream world. Keep it flexible, change it often, let it grow and expand as you do. But keep it ever in mind. Don't forget to live in the today. That's what we're here for. But live your life today as if that better world already existed - today. Don't wait for some theoretical future time to live it. Live it now. Enjoy it now. And someday, maybe in the not so distant future, you just may find yourself smack dab in the middle of your dream world. And discover you are not dreaming it, you are living it. And so are the people around you.

We wish you the best with this world, that world, and all the worlds in between.

THE HOMESTEAD SONG

Oh, we'll find us a homestead
We'll laugh in the wind
Find treasures in sunbeams
And life will begin

The chickens will crow
And the bulls will have calves
Our orange trees will blossom
Midst winter's cold blast

We'll need no such frou-frou
As money or cars
Our dreams will sustain us
We'll dance with the stars

The pantry will fill
To the brim and beyond
We'll feed all the neighbors
We'll swim in the pond

The black flies will tickle
The mosquitoes fill up
On flowers and moonbeams
And not on our blood

We'll learn what we need to
Our muscles will bulge
When we find our homestead
We'll manage it all

And when daylight surrounds me
And I wake from my dreams
You'll find it all happening
In spite what it seems

Bibliography

Following are books and sources we've found helpful, arranged by the chapters they pertain to. Some of the publications noted are not currently in print, however, you may be able to find a copy in a library or used book store.

1 - Homestead Dreams
2 - Beginnings and Bugs

Living the Good Life, and *Continuing the Good Life,* by Helen and Scott Nearing. Schocken Books, 1970.
Farming for Self-Sufficiency; Independence on a 5-Acre Farm, by John Seymour. Schocken Books, 1976.
Countryside and Small Stock Journal, W11564 Hwy 64, Withee, Wisconsin 54498. www.countrysidemag.com
BackHome magazine, PO Box 70, Hendersonville, North Carolina 28792.
Earth Quarterly, Box 23, Radium Springs, New Mexico 88054.
Small Farmer's Journal, PO Box 1627, Sisters, Oregon 97759.

3 - Walls, Roof and Windows - The Cabin in Born
4 - Where the Building Begins - Behind and Beyond
5 - Down to Earth, Under the Earth - The House

Architectural Graphic Standards, Fifth Ed., by Charles G. Ramsey/Harold R. Sleeper. Wiley & Sons, 1963.
Handmade Houses, A Guide to the Woodbutcher's Art, by Art Boericke/Barry Shapiro. Scrimshaw Press, 1973.
Low Cost, Energy-Efficient Shelter, For the Owner and Builder, by Eugene Eccli. Rodale Press, 1976.
From the Ground Up, by John Cole/Charles Wing. Atlantic Monthly Press/Little, Brown and Co., 1976.
Other Homes and Garbage, Designs for self-sufficient living, by Jim Leckie/Gil Masters/Harry Whitehouse/Lily Young. Sierra Club Books, 1975.
The $50 and Up Underground House Book, How to Design and Build Underground, by Mike Oehler. Mole Publishing, PO Box 270, Naples, Idaho 83847, 1979.
Underground Designs, by Malcolm Wells. Malcolm Wells, 1977.
Underground Plans Book-1, 8 Large-scale plans and details you can use in designing a house for your site, by Malcolm Wells/Sam Glenn-Wells. Malcolm Wells, 1980.
Building Construction Illustrated, by Francis Ching, Van Nostrand Reinhold Co., 1975.

The Timber Framing Book, by Stewart Elliott/Eugenie Wallas. Housesmiths Press, 1977.

The Owner Built Homestead, by Barbara and Ken Kern. Scribner, 1977.

How to Build a Low-Cost House of Stone, by Lewis & Sharon Watson. Stonehouse Publications, 1976.

Build Your Own Stone House, Using the easy, slipform method, by Karl & Sue Schwenke. Garden Way Publishing, 1975.

The Passive Solar Energy Book, A complete guide to passive solar home, greenhouse and building design, by Edward Mazria, Rodale Press, 1979.

The Solar Home Book, Heating, cooling and designing with the sun, by Bruce Anderson/Michael Riordan. Rodale Press, Cheshire Books, 1976.

6 - The Wind in the Mill and the Dance in the Water

Wind and Windspinners, by Michael Hackleman. Earthmind, 1974.

The Homebuilt, Wind-Generated Electricity Handbook, by Michael Hackleman. Peace Press 1976.

Other Homes and Garbage, Designs for self-sufficient living, by Jim Leckie/Gil Masters/Harry Whitehouse/Lily Young. Sierra Club Books, 1975.

7 - Touching Toes with Insects - The New Garden

How to Grow Vegetables and Fruits by the Organic Method, J. I. Rodale, Editor. Rodale Books, 1974.

The No Work Garden Book, by Ruth Stout and Richard Clemence. Rodale Press, 1971.

How to Have a Green Thumb Without an Aching Back, by Ruth Stout. Exposition PR of Florida, 1955.

Frost Dancing, Tips from a Northern Gardener, by Sue Robishaw. ManyTracks Publishing, 1998.

Down to Earth Vegetable Gardening Know How, by Dick Raymond. Garden Way Publishing, 1975.

Small-Scale Grain Raising, by Gene Logsdon. Rodale Press, 1977.

8 - The Joy of (Solar) Cooking

Home Power magazine, PO Box 520, Ashland, Oregon 97520. Issues #7, 12, 31, 37, 29, 43, 45.

Heaven's Flame Solar Cooker, by Joseph Radabaugh. Joseph Radabaugh, 1990. *Home Power* magazine.

10 - Breakfast Flowers and Luncheon Weeds
11 - Pickles and Eating Out . . . of the Garden

Mrs. Restino's Country Kitchen, by Susan Restino. Quick
    Fox 1976.
The Joy of Cooking, by Irma Rombauer, Marion Rombauer
    Becker. Bobbs-Merrill Company, 1975.
The No Work Garden Book, by Ruth Stout and Richard
    Clemence. Rodale Press, 1971.

12 - Power for the People - The Sun and Solar of Electricity

Home Power magazine, POB 520, Ashland, Oregon 97520,
    916-475-0830, hp@homepower.com, www.homepower.com
*People's Power Primer, Renewable Energy for the Technically
    Timid,* by Bob Dahse. Seldom Scene Productions, RR3 Box
    163-A, Winona MN 55987, 1999, 72 pgs, $11.00 ppd.
Electron Connection, POB 203, Hornbrook, California 96044,
    800-945-7587, econnect@snowcrest.net,
    www.electronconnection.net
Backwoods Solar Electric Systems, 1395 Rolling Thunder
    Ridge, Sandpoint, Idaho 83864, 208-263-4290,
    info@backwoodssolar.com, www.backwoodssolar.com
Alternative Energy Engineering, PO Box 339, Redway,
    California 95560, 800-777-6609, www.alt-energy.com
Midwest Renewable Energy Assoc., 7558 Deer Rd, Custer,
    Wisconsin 54423, 715-592-6595, mreainfo@wi-net.com,
    www.the-mrea.org

13 - Age Old Meets New Age at the Solar Food Dryer

*A Pantry Full of Sunshine, Energy Efficient Food Preservation
    Methods,* by Larisa Walk and Bob Dahse. Seldom Scene
    Productions, RR3 Box 163-A, Winona MN 55987, 1997,
    61 pgs, $8.50 ppd.
Home Power magazine, POB 520, Ashland, Oregon 97520.
    Issue #29.
Stocking Up, edited by Carol Hupping Stoner. Rodale Press,
    1977.
Putting Food By, by Ruth Hertzberg, Beatrice Vaughan,
    Janet Greene. The Stephen Greene Press, 1984.
Cambridge Wire Cloth Company, PO Box 399, Cambridge,
    Maryland 21613, 800-638-9560. [source for stainless steel
    screening, minimum order $100]

14 - The Days of Wine and Vinegar

Country Woodcraft, by Drew Langsner. Rodale Press, 1978.
    [bark berry baskets]

15 - To Save the Garden Seed

Seed to Seed, by Suzanne Ashworth. Seed Saver Publications, 1991.

Garden Seed Inventory, by Kent Whealy. Seed Saver Publications, 1995.

Fruit, Berry and Nut Inventory, by Kent Whealy and Steve Demuth. Seed Saver Publications, 1993.

Seed Savers Exchange: The First Ten Years, by Kent Whealy and Arllys Adelmann. Seed Saver Publications, 1986.

Growing Garden Seeds, A Manual for Gardeners and Small Farmers, by Robert Johnson, Jr., Johnny's Selected Seeds, Albion Maine 04910, 1983.

Frost Dancing, Tips from a Northern Gardener, by Sue Robishaw. ManyTracks Publishing, 1998.

Seed Savers Yearbook, Seed Savers Summer Edition, and *Seed Savers Harvest Edition* are annual publications to members of the Seed Savers Exchange.

For more information on the *Seed Savers Exchange* and their publications send $1 to: *Seed Savers Exchange,* 3076 North Winn Road, Decorah, Iowa 52101.

OPEN POLLINATED SEED SOURCES

Bountiful Gardens, 5798 Ridgewood Rd, Willits, CA 95490

High Altitude Gardens, PO Box 1048, Hailey, ID 83333

Garden City Seeds, 1324 Red Crow Rd, Victor, MT 59875

Prairie Grown Garden Seeds, Box 118, Cochin, SK S0M0L0

Fisher's Seeds, PO Box 236, Belgrade, MT 59714

Southern Exposure Seed Exchange, POB 170, Earlysville, VA 22936

Johnny's Selected Seeds, RR 1 Box 2580, Albion, ME 04910

Seeds Blum, Idaho City Stage, Boise, ID 83706

Abundant Life Seed, PO Box 772, Port Townsend, WA 98368

Good Seed Co., Star Rt Box 73A, Oroville, WA 98844

The Cook's Garden., PO Box 535, Londonderry, VT 05148

Turtle Tree Seed, 5569 North Co Rd 29, Loveland, CO 80538

Gleckler's Seedmen, Metamora, OH 43540

Irish Eyes with Garlic, PO Box 307, Ellensburg, WA 98926

Filaree Farm (garlic), 182 Conconully Hwy, Okanogan, WA 98840

Bear Creek Nursery, PO Box 411, Northport, WA 99157

16 - The Hunt for Red Potato and the Path to the Packed Pantry

Ronniger's Seed Potatoes, Star Rt, Moyie Springs, Idaho 83845. [catalog].

Stocking Up, edited by Carol Hupping Stoner. Rodale Press, 1977.

Putting Food By, by Ruth Hertzberg, Beatrice Vaughan, Janet Greene. The Stephen Greene Press, 1984.

17 - Cooking and Baking the ManyTracks Way

Mrs. Restino's Country Kitchen, by Susan Restino. Quick Fox, 1976.

The Joy of Cooking, by Irma Rombauer and Marion Rombauer Becker. Bobbs-Merrill Company, 1975.

18 - For Love and a Little Money - Life and Livelihood

Small Time Operator, by Bernard Kamoroff. Bell Spring Publishing, 1996.

19 - Maple Syrup, Sweet Success

The Maple Sugar Book, by Helen & Scott Nearing. Schocken Books, 1970.

20 - Great Expectations, Small Greenhouse

Winter Flowers in Greenhouse and Sun-heated Pit, by Kathryn Taylor and Edith Gregg. Charles Scribner's Sons, 1969.

The Solar Greenhouse Book, edited by James McCullagh. Rodale Press, 1978.

Building and Using our Sun-Heated Greenhouse, by Helen Nearing. Garden Way Publishing, 1977.

Other

Herbal Handbook for Farm and Stable, by Juliette de Bairacli Levy. Rodale Press, 1976.

The Homesteader's Handbook to Raising Small Livestock, by Jerome D. Belanger. Rodale Press, 1974.

The Contrary Farmer, by Gene Logsdon. Chelsea Green Publishing, 1995.

Build It Better Yourself, by the editors of Organic Gardening and Farming. Rodale Press, 1977.

The Place Called Attar - A Loving Story of the Earth and one family who lived in harmony with Her, by J. D. Belanger. Countryside Publications, 1992.

ManyTracks - Sue Robishaw, Steve Schmeck

Come visit us on-line at **http://www.up.net/~manytrac**

Write to us via email: manytrac@up.net

or postal mail: Rt 1 Box 52 B, Cooks, Michigan 49817

Index

Colophon

Designed and edited by
Sue Robishaw and Steve Schmeck
Published by ManyTracks Publishing
Composed and set in Adobe PageMaker
Text font is New Century Schoolbook Roman, 11 pt
Graphics fonts are Arial and Courier New
Accent font is Albertville
Printed and bound by Patterson Printing,
Benton Harbor, Michigan
Soyink used throughout
Text paper is Fraser Papers Halopaque
100% recycled, acid-free, 60# text
The entire book, except for the printing, was created in
an alternative energy environment

About the Author

Sue Robishaw lives with her husband, Steve Schmeck, two cats, and much wildlife in the back woods of Michigan's Upper Peninsula. There they have enjoyed a sustainable, self-reliant lifestyle for more than twenty years, and are active proponents of hands-on living.

An avid gardener and seed saver, Robishaw is a frequent speaker and workshop presenter. Her articles have been published in a number of magazines and publications; she is the author of *The Last Lamp*, a suspense novel of the 21st century; *Carlos's ma's Friends*, a contemporary novella with an urban setting; *Frost Dancing—Tips From a Northern Gardener*, a booklet full of gardening ideas; and *Rosita and Sian Search for a Great Work of Art*, a humorous allegory on creativity, for all ages.

$$\mathcal{MT}$$
ManyTracks

ManyTracks Publishing

ManyTracks Publishing is a publisher of quality books printed on 100% recycled paper. All of the books are created in an alternative energy environment with power from the sun, and are available from bookstores, libraries, catalogues, or from ManyTracks Publishing. Send for a current catalog, or visit the ManyTracks web site at:

www.up.net/~manytrac

ManyTracks Publishing
R 1 Box 52B
Cooks MI 49817

manytrac@up.net